# Scalable Network Monitoring in High Speed Networks

Baek-Young Choi · Zhi-Li Zhang
David Hung-Chang Du

# Scalable Network Monitoring in High Speed Networks

 Springer

Baek-Young Choi
Department of Computer Science
 and Electrical Engineering
University of Missouri – Kansas City
546 Flarsheim Hall
Rockhill Rd. 5110
64110 Kansas City MO
USA
choiby@umkc.edu

Zhi-Li Zhang
Department of Computer Science
 and Engineering
University of Minnesota
4-192 EE/CS Building
Union Street SE 200
55416 Minneapolis MN
USA

David Hung-Chang Du
Department of Computer Science
 and Engineering
University of Minnesota
4-192 EE/CS Building
Union Street SE 200
55455 Minneapolis MN
USA

ISBN 978-1-4899-8563-7        ISBN 978-1-4614-0119-3 (eBook)
DOI 10.1007/978-1-4614-0119-3
Springer New York Dordrecht Heidelberg London

Printed on acid-free paper

Springer is part of Springer Science+Business Media (www.springer.com)

*Dedicated to our families*

# Preface

*Network monitoring* serves as basis for a wide scope of network operations, engineering and management. Precise network monitoring involves inspecting every packet traversing in a network. However, this is infeasible in today's and future high-speed networks, due to significant overheads of processing, storing, and transferring measured data. Therefore, *scalable* network monitoring techniques are in urgent need. This book addresses the scalability issue of network monitoring from both traffic and performance perspectives.

On scalable traffic monitoring, we present sampling techniques for *total load* and *flow* measurement. In order to develop accurate and efficient measurement schemes, we study various aspects of traffic characteristics and their impacts on packet sampling. We find that static sampling does not adjust itself to dynamic traffic conditions, yielding often erroneous estimations or excessive oversampling. We develop the *adaptive random sampling technique* for total load estimation, that determines the sampling probability adaptively according to traffic condition. We then enhance the adaptive sampling technique to measure traffic in flow level. Flow measurement is a particularly challenging problem, since flows arrive at random times, stay for random durations, and their rates fluctuate over time. Those characteristics make it hard to decide a sampling interval where sampling probability is adapted, and to define a large flow pragmatically. Through a stratified approach, we estimate large flows accurately, regardless of their arrival times, durations, or the rate variabilities during their life times.

On practical performance monitoring, we investigate issues around network *delay*. We first perform a detailed *analysis of point-to-point packet delay in an operational tier-1 network*. Through a systematic methodology and careful analysis, we identify the major factors that contribute to point-to-point delay, and characterize their effect on the network delay performance. Next, we identify *high quantile* as a meaningful metric in summarizing a delay distribution. Then, we develop an *active sampling scheme* to estimate a high quantile of a delay distribution with bounded error. We finally show that active probing is the most scalable approach to delay measurement. The validation of our proposed schemes and the analysis of network traffic and performance presented in this book are conducted with *real operational network traces*.

Many people have helped us with various expertise needed for this monograph. We especially thank Drs. Christophe Diot, Sue Moon, Rene Cruz, Dina Papagiannaki, and Jaesung Park for their invaluable guidance and insights. The collaboration with them gave a definite direction and shape to this work.

# Contents

# List of Figures

# Chapter 1
# Overview

## 1.1 Motivation and Approach

The Internet has grown dramatically over the past several years, in terms of the amount of traffic and the number of usages as well as connectivity. With this rapid growth, the Internet is used for increasingly diverse and demanding purposes, which lead to frequent changes in network status, yet raised performance requirements. Therefore, it is crucial to monitor the network, in order to understand its behavior and to react appropriately, and furthermore, to efficiently design and provision the future network.

Important network traffic and performance parameters to monitor include the following.

The amount of traffic at various granularities are necessary for diverse measurement applications, for instance:

- Total traffic load: The total amount of traffic observed on a link is the basic indicator for network activity.
- Flow volume: Flow volume is the amount of traffic used by a certain user, application, or protocol. Flow level measurement can be used for usage profiling, user accounting, load balancing, and the basic information from the link level for the network wide traffic matrix.
- Traffic matrix: Traffic matrix is a matrix that gives a network-wide view of traffic flowing between any source and destination.

Examples of network performance parameters are:

- Delay: Latency experienced by packets within a network indicates the canonical network performance in terms of configuration and congestion of the network.
- Jitter: Delay variation presents how stable the network is as perceived by users.
- Loss: Occurrence of packet drops by network elements that denote how often packets are not delivered through a network.

Precise traffic and performance measurement involve inspecting every packet traversing in a network. However, they incur significant overhead with today's and future high speed Internet backbone links.

- Processing overhead: Many kinds of monitoring, such as flow measurement and on-line analysis are performed by software whose processing speed cannot match the line speed. Furthermore, most external measurement devices [1, 13] do not keep up with packet rates in high speed links.
- Storage requirement: Disks are quickly exhausted due to a huge number of packets causing discontinuity of data.[1] In case of flow measurement, the cache memory in the router, which keeps flow records, can overflow when a large number of flows are arriving concurrently.
- Bandwidth consumption: Large bandwidths are consumed, when measured data, either raw packets or some level of aggregated statistics, should be transferred to an operational center for the purpose of record or further processing.

Scalable network monitoring techniques are thus, greatly demanded. This book addresses the scalability issue from both traffic and performance perspectives. *Sampling* techniques may be used as a way to limit the measurement overhead as proposed in PSAMP [15] and IPFIX [12] Internet Working Groups. While passive measurement is used for traffic monitoring, often packets are injected for the purpose of performance measurement [14], and the performance of probe packets are used to infer the network performance. Then, the active probing essentially performs as active sampling, under the following case. First, the amount of probe packets should be negligible as compared to the total traffic so that it does not perturb the performance it measures. Second, the performance of probe packets should well represent the performance of regular traffic.

The foremost and fundamental issue regarding sampling is its *accuracy* or representativeness. This is especially pertinent on the Internet, where network status is known to fluctuate dynamically and frequently. Inaccurate packet sampling not only defeats the purpose of network measurement and monitoring, but worse, can lead to wrong decisions by network operators. An important related concern is the *efficiency* of packet sampling. Excessive oversampling should also be avoided for the measurement solution to be scalable. Therefore, it is important to *control the accuracy* of estimation in order to *balance the trade-off between the utility and overhead of measurement*. This book is centered in designing monitoring schemes to produce representative and accurate measurement of network traffic and performance, while minimizing measurement overhead.

## 1.2 Main Contributions

In this book, we have addressed the scalability issue of network monitoring from both traffic and performance perspectives. We first investigated the characteristics

---

[1] For example, 5 hours of OC-48 link trace with 30% to 40% utilization takes larger than 300 GB.

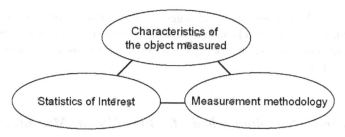

**Fig. 1.1** Our approach: Understanding and exploiting the relation of the above three aspects.

of network traffic and performance parameters, then developed sampling techniques to measure them accurately and efficiently. Figure 1.1 illustrates our approach to network monitoring. The characteristics of the object measured, metric or statistics of interest, and measurement methodology are tightly coupled. In order to produce measurements accurately and efficiently, a measurement scheme should be designed by taking the object characteristic and the measured statistic into consideration. Thus, it is important to understand the characteristics of network parameters of interest that are advanced to develop measurement methods. Together with the applications of measurement, the characteristics of network traffic and performance may provide insights on what can be and what should be measured with a measurement scheme, as well as a reasonable and meaningful measurement metric. We apply and enhance sampling theory to measure the metric of the network parameter to ensure the required accuracy, while avoiding unnecessary oversampling for the measurement solution to be scalable. Below we give an overview of each part of the book.

### 1.2.1 Adaptive Sampling Scheme for Traffic Load Measurement

Monitoring the current level of network activity is the first step towards traffic engineering. Thus, it is essential to measure traffic load accurately and in a timely manner. In estimating traffic load with packet sampling, we address the problem of bounding sampling errors within a pre-specified tolerance level. We find that static sampling does not adjust itself to a dynamic traffic condition, often yielding an erroneous estimation or excessive oversampling. We derive a relationship between the number of packet samples, the accuracy of load estimation and the squared coefficient of variation of packet size distribution. Based on this relationship, we design an adaptive random sampling technique that determines the minimum sampling probability that adapts according to traffic dynamics [6, 8, 9, 11]. Using real network traffic traces, we show that the proposed adaptive random sampling technique indeed produces the desired accuracy, while also yielding significant reduction in the amount of traffic samples that are simple to implement. We also apply the proposed sampling technique for automatic load change detection, as an application of on-line traffic analysis. Timely detection of changes in traffic is critical for initiating appropriate traffic engineering mechanisms. We investigate the impact of sampling errors on the performance of load change detection, and illustrate how crucial it is to have accurate estimations.

### 1.2.2 Adaptive Sampling Scheme for Flow Volume Measurement

We analyze and explore flow-level traffic characteristics to understand the challenges and the theoretical and practical issues involved in sampling. Motivated by the phenomenon that a small portion of large flow accounts for the most of total traffic, we endeavor to estimate flows accurately by focusing only on large flows. We enhance and develop the adaptive, stratified packet sampling technique for flow measurement [7, 10, 11], considering the dynamic nature of flows such as random arrival and duration, as well as rate variability within a flow duration. Through theoretical studies and experiments, we demonstrate that the proposed sampling technique provides an unbiased estimation of flow size with controllable error bound, in terms of both packet and byte counts for elephant flows, while avoiding excessive oversampling.

### 1.2.3 Analysis of Point-to-Point of Delay in an Operational Backbone

We perform a detailed analysis of point-to-point packet delay in an operational tier-1 network [4, 5]. The point-to-point delay is the time between a packet entering a router in one PoP (an ingress point) and its leaving a router in another PoP (an egress point). It measures the one-way delay experienced by packets from an ingress point to an egress point across a carrier network and provides the most basic information regarding the delay performance of the carrier network. Using packet traces captured in the operational network, we obtain precise point-to-point packet delay measurements and analyze the various factors affecting them. Through a simple, step-by-step, systematic methodology and careful data analysis, we identify the major network factors that contribute to point-to-point packet delay and characterize their effect on the network delay performance. Our findings are 1) delay distributions vary greatly in shape, depending on the path and link utilization; 2) after constant factors dependent only on the path and packet size are removed, the 99th percentile variable delay remains under 1 ms over several hops and under link utilization below 90% on a bottleneck; 3) a very small number of packets experience very large delays in short bursts.

### 1.2.4 Practical Delay Measurement for an ISP

We find a meaningful representation of network delay performance in high quantiles, and use a random sampling technique to estimate them accurately [2, 3]. Our contribution in this work is three-fold. First, by analyzing the delay measurement data from an operational network (Sprint US backbone network), we identify high-quantiles {0.95-0.99} as the most meaningful delay metrics that best reflect the delay experienced by most of packets in an operational network, and suggest 10-30 minute time scale as an appropriate interval for estimating the high-quantile delay metrics. The high-quantile delay metrics estimated over such a time interval provide a best representative picture of the network delay performance that captures the major changes and trends, while they are less sensitive to transient events, and outliers. Second, we design and develop an active probing method for estimating high-quantile delay metrics. The novel feature of our proposed method is that it uses the minimum number of samples needed to bound the error of quantile estimation within a prescribed accuracy, thereby reducing the measurement overheads of active probing. We finally compare the network wide overhead of active probing and passive sampling for delays, and show that active probing is the most scalable approach to delay measurement.

## 1.3 Organization

The rest of this book is organized as follows. The book is divided into two parts. In the first part of this book, we focus on *network traffic*. We study characteristics of Internet traffic, and develop sampling mechanisms for traffic monitoring. In Chapter 2, we study the characteristics of traffic load, and develop an adaptive random sampling technique to estimate load In Chapter 3, we explore the issue of traffic flow measurement, and enhance the adaptive random sampling technique to measure flows efficiently and accurately. The second part of the book is dedicated to network *delay* which is the most important *network performance* parameter. In Chapter 4, we present a step-by-step analysis of point-to-point delay from an operational tier-1 backbone network. Based on the insight gained from the analysis, in Chapter 5, we develop a practical delay measurement scheme for an ISP. We validate all the techniques proposed in this book using real network traces. Finally, in Chapter 6, we summarize out contributions and discuss possible directions for future research.

## References

1. B. Carpenter. Middleboxes: Taxonomy and issues. Internet Engineering Task Force Request for Comments: 3234, February 2002.

2. B.-Y. Choi, S. Moon, R. Cruz, Z.-L. Zhang, and C. Diot. Practical Delay Measurement Methodology in ISPs. In *Proceedings of ACM CoNext (Nominated for best paper award)*, Toulouse, France, Oct. 2005.

3. B.-Y. Choi, S. Moon, R. Cruz, Z.-L. Zhang, and C. Diot. Quantile Sampling for Practical Delay Monitoring in Internet Backbone Networks. *Elsevier Journal of Computer Networks*, 51(10):2701–2716, Jul. 2007.

4. B.-Y. Choi, S. Moon, Z.-L. Zhang, K. Papagiannaki, and C. Diot. Analysis of Point-to-Point Packet Delay in an Operational Network. In *Proceedings of IEEE INFOCOM (Nominated for best paper award)*, Hong Kong, Mar. 2004.

5. B.-Y. Choi, S. Moon, Z.-L. Zhang, K. Papagiannaki, and C. Diot. Analysis of Point-To-Point Packet Delay In an Operational Network. *Elsevier Journal of Computer Networks*, 51(13):3812–3827, Sep. 2007.

6. B.-Y. Choi, J. Park, and Z.-L. Zhang. Adaptive Random Sampling for Load Change Detection. Technical report, University of Minnesota, 2001.

7. B.-Y. Choi, J. Park, and Z.-L. Zhang. Adaptive Packet Sampling for Flow Volume Measurement. *ACM SIGCOMM Computer Communication Review*, 32(3), 2002.

8. B.-Y. Choi, J. Park, and Z.-L. Zhang. Adaptive Random Sampling for Load Change Detection. *ACM SIGMETRICS Performance Evaluation Review*, 30(1):272 – 273, Jun. 2002.

9. B.-Y. Choi, J. Park, and Z.-L. Zhang. Adaptive random sampling for traffic load measurement. In *Proceedings of ICC*, Anchorage, AK, May 2003.

10. B.-Y. Choi, J. Park, and Z.-L. Zhang. Adaptive Packet Sampling for Accurate and Scalable Flow Measurement. In *Proceedings of IEEE Global Internet Symposium (Globecom)*, Dallas, TX, Nov/Dec. 2004.

11. B.-Y. Choi and Z.-L. Zhang. Adaptive Random Sampling for Traffic Volume Measurement. *Springer Journal of Telecommunication Systems (TS)*, 34(1–2), Feb. 2007.

12. IPFIX. Internet Engineering Task Force, IP Flow Information Export. *http://www.ietf.org/html.charters/ipfix-charter.html*.

13. Sprint ATL IPMon project. *http://ipmon.sprint.com*.

14. IPPM. Internet Engineering Task Force, IP Performance Metric charter. *http://www.ietf.org/html.charters/ippm-charter.html*.

15. PSAMP. Internet Engineering Task Force Packet Sampling working group. *https://ops.ietf.org/lists/psamp*.

# Part I
# Scalable Traffic Monitoring with Packet Sampling

# Chapter 2
# Load Characterization and Measurement

**Abstract** Timely detection of changes in traffic is critical for initiating appropriate traffic engineering mechanisms. Accurate measurement of traffic is an essential step towards change detection and traffic engineering. However, *precise* traffic measurement involves inspecting *every* packet traversing a link, resulting in significant overhead, particularly on routers with high speed links. *Sampling* techniques for traffic load *estimation* are proposed as a way to limit the measurement overhead. Since the efficacy of change detection depends on the *accuracy* of traffic load estimation, it is necessary to *control* error in estimation due to sampling. In this paper, we address the problem of *bounding* sampling error within a pre-specified tolerance level. We derive a relationship between the number of packet samples, the accuracy of load estimation and the squared coefficient of variation of packet size distribution. Based on this relationship, we propose an *adaptive random sampling* technique that determines the *minimum* sampling probability adaptively according to traffic dynamics. Using real network traffic traces, we show that the proposed adaptive random sampling technique indeed produces the desired accuracy, while also yielding significant reduction in the amount of traffic samples, yet simple to implement. We also investigate the impact of sampling errors on the performance of load change detection.

**Key words:** adaptive random sampling, traffic measurement, packet trace, estimation accuracy, traffic load, change detection

## 2.1 Introduction

With the rapid growth of the Internet, traffic engineering has become an important mechanism to reduce network congestion and meet various user demands. Measurement of network traffic load is crucial for configuring, managing, pricing, policing, and engineering the network. Network traffic may fluctuate frequently and often unexpectedly for various reasons such as transitions in user behavior, deployment of new applications, changes in routing policies, or failure of network elements. It is a daunting task for network administrators to manually tune the network configuration to accommodate the traffic dynamics. Thus, there is a need for automatic tools that enable intelligent control and management of high speed networks. Clearly, *accurate* measurement of traffic load is a prerequisite to on-line traffic analysis tools.

On-line traffic analysis tools are only implemented on a separate measurement-box. It is because traffic analysis is not a part of router functionality in the current routers, and it is easier to develop and augment analysis functions separately from a router's built-in software.

The Simple Network Management Protocol (SNMP) [21] provides router status information such as interfaces, routing tables, and protocol states defined by Management Information Base (MIB). Hosts or routers, as SNMP agents, can make requests to pull out MIB variables. In backbone routers, however, SNMP polling interval cannot be set short, since the router's performance should not be degraded replying SNMP requests, and bandwidth should not be waisted too much for SNMP messages. Also, SNMP is processed at a low priority in a router. Thus, often SNMP is not flexible, and the data is not readily available for timely analysis on-the-fly.

For out-of-the-box (or off-board) measurement systems, inspecting every single packet that traverses a link, however, is extremely costly. It may not keep up with today's high-speed of links. With off-board measurement devices, huge volumes of data are generated that can quickly exhaust storage space.

Sampling techniques may provide timely information economically, in particular if on-line analysis of the data is needed. However, sampling inevitably introduces errors in the traffic load estimation. Such errors may *adversely* affect the measurement applications.

In this chapter, we develop an *adaptive random sampling* technique for traffic load measurement. Our adaptive random sampling technique differs from existing sampling techniques for traffic measurement in that it yields *bounded* sampling errors *within a pre-specified error tolerance level*. Such error bounds are important in reducing the "noise" in the sampled traffic measurements. Furthermore, the pre-specified error tolerance level allows us to control the performance of on-line traffic analysis algorithms as well as the amount of packets sampled. This chapter is devoted to the analysis and verification of the proposed adaptive random sampling technique for traffic load measurement. We also study the impact of sampling errors on the performance of on-line traffic analysis - load change detection. Our contributions are summarized as follows.

We observe that sampling errors in estimating traffic load arises from the dynamics of packet sizes and counts, and these traffic parameters vary over time. Consequently, *static* sampling (i.e., with a fixed sampling rate) cannot guarantee errors within a given error tolerance level. From our analysis, we find that the *minimum* required number of samples to bound a sampling error within a given tolerance level is proportional to the squared coefficient of variation (*SCV*) of packet size distribution. Using this relationship, we propose an adaptive random sampling technique that determines the (minimum) sampling probability adaptively based on the *SCV* of packet size distribution and the packet count. More specifically, time is divided into (non-overlapping) observation periods (referred to as (time) blocks), and packets are sampled in each observation period. At the end of each block, in addition to estimating the traffic volume of the block, the *SCV* of packet size distribution and the packet count of the block are calculated using the traffic samples. These traffic parameters are used to predict the *SCV* of packet size distribution and the packet count of the next block, using an *Auto-Regressive* (AR) model. The sampling probability for the next block is then determined based on these predicted values and the given error tolerance level. The procedure is depicted in Figure 2.1. Through analysis, we quantify the estimation and prediction errors introduced by our sampling technique, and devise mechanisms to control their impact on the traffic load estimation. Using real network traffic traces, we show that the proposed adaptive random sampling technique indeed produces the desired accuracy, while at the same time yielding significant reduction in the amount of traffic samples. Using the time series of the estimated traffic load, we present a non-parametric on-line change point detection algorithm based on singular value spectrum analysis. The algorithm finds nonstationarities in traffic loads at some larger, configurable, operational time scale using sampled measurements obtained at each (smaller time-scale) observation period. The basic approach is depicted in Figure 2.2. We investigate the impact of sampling errors on the performance of this load change detection algorithm using real network traffic traces.

The remainder of the chapter is structured as follows. In Section 2.2, we outline related works on traffic measurement and analysis, as well as several sampling methods that have been used and studied for various applications. In Section 2.3, we formally state the problem addressed in this chapter. In Section 2.4, the adaptive random sampling technique is described and analyzed. Experimental results with real network traffic traces are presented in Section 2.5. We present the change point detection algorithm with sampled measurement in Section 2.6. Section 2.7 summarizes the chapter.

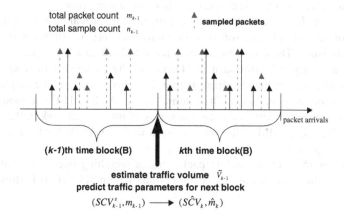

**Fig. 2.1** Adaptive random sampling.

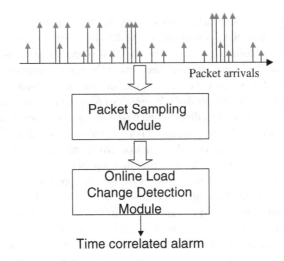

**Fig. 2.2** Load change detection system model.

## 2.2 Related Works

In this section, we describe a few traffic measurement infrastructures and several sampling methods that have been applied or proposed in the context of a communication network.

Cisco's Internetworking Operating System (IOS) has a passive monitoring feature called NetFlow [6]. NetFlow allows a router to collect flow level traffic statistics such as flow packet count, byte count, start time, and ending time.

*CAIDA's CoralReef* is a comprehensive software suite providing a set of drivers, libraries, utilities, and analysis software for passive network measurement of workload characteristics. These reports characterize workload on a high-speed link between the University of California at San Diego and the commodity Internet.

A few traffic measurement infrastructures have been built, as the importance of traffic measurement increases. The Internet Traffic Archive [14] is a moderated repository to support widespread access to traces of Internet network traffic, sponsored by ACM SIGCOMM [20]. Web traces, as well as LAN/WAN traces, can be found on the repository.

The CAIDA [5] provides public packet traces including ones collected on a number of monitors located at aggregation points within NSF High Performance

Connections (HPC), the NSF/MCI very high performance Backbone Network Service (vBNS), and Internet2 Abilene[1] sites.

Packet traces from a set of selected links inside the Sprint backbone network are collected, and some of the statistics are available from [13], even though the traces themselves are not public.

Our study on traffic characteristics and the validation of the proposed schemes are carried on using traces from both [5] and [13].

Statistical sampling of network traffic was first used in [7] for measuring traffic on the NSFNET backbone in the early 1990's. Claffy *et al.* evaluated classical events and time-driven static sampling methods to estimate statistics of distributions of packet size and inter-arrival time, and observed that the event driven outperforms the time driven approach.

In [8], the authors applied a random packet sampling to evaluate the ATM end-to-end QoS such as the cell transfer delay.

The hash-based sampling proposed in [9] employs the same hashing function at all links in a network to sample the same set of packets at different links, and in order to infer statistics on the spatial relations of the network traffic.

A size-dependent *flow* sampling method proposed in [10] addresses the issue of reducing the bandwidth needed for the transmission of traffic measurement to a management center for later analysis. For the purpose of usage-based charging, flows are probabilistically sampled depending on their sizes, assuming flow statistics are known a priori.

In [11], a probabilistic packet sampling method is used to identify large byte count flows. Once a packet from a flow is sampled or identified, all the subsequent packets belonging to the flows are sampled. It underestimates flow sizes, since the packets arriving before the flow is determined to be a large flow are not counted. Also, in order to make sampling decision, every packet should be classified into a flow, and checked if the flow is already in the cache or not. Thus, it increases per-packet processing overhead, to limit the size of flow cache.

The study in [15] presents an algorithm to bound the flow packet count estimation error of the top $k$ largest flows. It performs on a fixed measurement interval, from which one set of top largest flows are estimated.

Our work differs from the above, in that we use packet sampling to estimate traffic volume in terms of both packet and byte counts within a *pre-specified error bound under dynamic traffic conditions*.

## 2.3 Sampling Problem for Traffic Load Measurement

In this section, we first formulate the sampling problem for traffic load measurement. Monitored traffic load will be used for time-series analysis such as detecting

---

[1] Abilene is an advanced backbone network that connects regional network aggregation points, to support the work of Internet2 universities as they develop advanced Internet applications [1].

abrupt changes in traffic loads. We then derive a lower bound on the number of samples needed to estimate the traffic load accurately within a given tolerance level. Based on this, we determine the sampling probability that is *optimal* in the sense that it guarantees the given accuracy with the minimum number of samples. The optimal sampling probability depends on both the number of packets and the variation in their sizes in an observation period. We see that the network traffic fluctuates significantly over time in terms of both the number of packets and their sizes. Hence, the optimal sampling probability also varies over time. This suggests that *static* sampling with *fixed* sampling probability may result in either erroneous undersampling or unnecessary oversampling. In other words, static sampling cannot capture the traffic dynamics accurately or efficiently. This motivates us to develop an *adaptive* random sampling technique that attempts to minimize the sampling frequency while ensuring that the sampling error is bounded.

### 2.3.1 Bounding Sampling Errors in Traffic Load Estimation

Traffic load is the sum of the sizes of packets arriving during a certain time interval. Thus, traffic load is determined by both the number of packets and their sizes. In determining the traffic load, the variability of packet sizes is often overlooked and only packet count is considered. However, average packet size plays an important role in estimating the traffic load. Consider, for example, two network traffic traces captured at the University of Auckland to a US link. The time series plots of the traffic loads of the two traces ($\Pi_4$ and $\Pi_5$ in Table 2.2) are shown in the top row of both Figure 2.3(a) and 2.3(b). The plots in the middle row show the average packet sizes over time, while the plots in the bottom row show the packet counts over time. From Figure 2.3(a), we see that the increase in the traffic load around 1000 sec is due to the increase in the packet size rather than the packet count. On the other hand, the abrupt increase in the packet count near $2 \times 10^4$ sec in Figure 2.3(b) does not lead to any increase in the traffic load, since the packet sizes at the time are extremely small. These examples illustrate that the variation in packet sizes is an important factor in estimating the traffic load using sampling. In fact, we will later show that the variation in packet sizes is the key factor in determining the sampling rate and for controlling the accuracy of load estimation.

The reason that we highlight the factors that affect the traffic load estimation using sampling is that an *accurate* estimation of traffic load is crucial in measurement applications such as on-line change detection. For change point detection, the series of the estimated traffic loads must retain the *change* or *stability* of the original traffic. Significant sampling errors in traffic load estimation can distort the original "signal" and lead to *false alerts* that may adversely affect the performance of networks, for instance, if they inadvertently trigger inappropriate traffic engineering mechanisms. Hence, quantifying and bounding sampling errors is critical in applying sampling techniques to traffic load estimation for the purpose of load change detection.

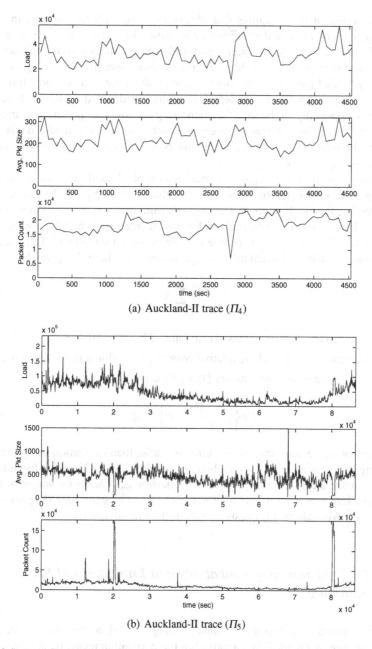

(a) Auckland-II trace ($\Pi_4$)

(b) Auckland-II trace ($\Pi_5$)

**Fig. 2.3** Impact of packet size and packet count on traffic load.

Time series analysis requires that observations be uniformly spaced in time. Packet arrivals at a link in the Internet are by nature irregularly spaced in time, and so are the packet samples. To obtain a uniformly spaced time series, traffic loads can be estimated from packets sampled during (non-overlapping) observation periods of fixed length (see Figure 2.1). We refer to an observation period as a (load estimation) *time block*, or simply *block*. The length of a block is denoted by $B$, which can be configured depending on the specific engineering purposes. To preserve the trend of the original traffic load, the sampling error in each block must be bounded quantitatively. In the following, we formally state the problem of bounding sampling errors in traffic load estimation.

Assume that there are $m$ packets arriving in a block, and let $X_i$ be the size of the $i$th packet. Hence, the traffic load of this block is $V = \sum_{i=i}^{m} X_i$. To estimate the traffic load of the block, suppose we *randomly* sample $n$, $1 \leq n \leq m$, packets out of the $m$ packets. In other words, each packet has an equal probability $p = n/m$ to be sampled. Let $\hat{X}_j$, $j = 1, 2, \ldots, n$, denote the size of the $j$th sampled packet. Then the traffic load $V$ can be estimated by $\hat{V}$ using the samples, where $\hat{V}$ is given by

$$\hat{V} = \frac{m}{n} \sum_{j=1}^{n} \hat{X}_j \tag{2.1}$$

It can be shown that $\hat{V}$ is an unbiased estimator of $V$, i.e., $E[\hat{V}] - V = 0$.

Our objective is to bound the relative error $\left| \frac{\hat{V} - V}{V} \right|$ within a *prescribed* error tolerance level given by two parameters $\{\eta, \varepsilon\}$ ($0 < \eta, \varepsilon < 1$), i.e.,

$$Pr \left\{ \left| \frac{\hat{V} - V}{V} \right| > \varepsilon \right\} \leq \eta. \tag{2.2}$$

In other words, we want the relative error in traffic load estimation using random sampling to be bounded by $\varepsilon$ with a high probability $1 - \eta$. Given this formulation of the bounded error sampling problem, the question is, '*what is the minimum number of packets that must be sampled randomly so as to guarantee the prescribed accuracy?*'. We address this question in the following subsection.

### 2.3.2 Optimal Sampling Probability and Limitations of Static Sampling

From the central limit theorem of *random samples*[2] [3], as the sample size $n \to \infty$, the average of sampled data approaches the population mean, regardless of the distribution of the population. Thus (2.2) can be rewritten as follows:

---

[2] The requirement that samples be i.i.d (independent and identically distributed) is achieved by *random* sampling from the same population.

$$Pr\left\{\left|\frac{\hat{V}-V}{V}\right|>\varepsilon\right\} = Pr\left\{\frac{\sqrt{n}}{\sigma}\left|\frac{1}{n}\sum_{i=1}^{n}X_i-\mu\right|>\frac{\varepsilon\mu\sqrt{n}}{\sigma}\right\} \tag{2.3}$$

$$\approx 2\left(1-\Phi\left(\frac{\varepsilon\mu\sqrt{n}}{\sigma}\right)\right)\leq\eta,$$

where $\mu$ and $\sigma$ are, respectively, the population mean and standard deviation of the packet size distribution in a block, and $\Phi(\cdot)$ is the cumulative distribution function (c.d.f) of the standard normal distribution (i.e., $N(0,1)$). Hence, to satisfy the given error tolerance level, the required number of packet samples must satisfy

$$n\geq n^* = \left(\frac{\Phi^{-1}(1-\eta/2)}{\varepsilon}\cdot\frac{\sigma}{\mu}\right)^2 = z_p\cdot S \tag{2.4}$$

where $z_p = \left(\frac{\Phi^{-1}(1-\eta/2)}{\varepsilon}\right)^2$ and $S = (\sigma/\mu)^2$ is the squared coefficient of variance (*SCV*) of the packet size distribution in a block. Eq. (2.4) *concisely relates* the minimum number of packet samples to the estimation accuracy and the variability in packet sizes. In particular, it states the minimum required number of packet samples, $n^*$, is *linearly* proportional to the squared coefficient of variance, $S$, of the packet size distribution in a block.

From (2.4) we conclude that the *optimal* sampling probability, $p^*$, that samples the minimum required number of packets in a block, is given by

$$p^* = \frac{n^*}{m}. \tag{2.5}$$

Hence, to attain the prescribed sampling accuracy $\{\eta,\varepsilon\}$, packets in a block must be sampled randomly with a probability of at least $p^*$. Note that, to determine the optimal sampling probability $p^*$, we need to know the *actual SCV* of the packet size distribution and the packet count $m$ in a block. Unfortunately, in practice, these traffic parameters of a block are *unknown* to us at the time the sampling probability for the block must be determined. To circumvent this problem, in Section 2.4 we develop an AR (Auto-Regressive) model to predict these parameters of a block based on past sampled measurements of previous blocks. Before we proceed to present this model, we would like to conclude this section by discussing the limitations of *static* sampling.

Static sampling techniques such as "one-out-of-N" sampling are commonly employed in routers, as they are simple to implement. For example, Cisco's Sampled NetFlow [18] introduced in IOS 12.0(3)T samples one packet out of every $N$ IP packets for flow statistics.[3] More generally, the static *random* sampling technique randomly samples a packet with a *fixed* probability. Both techniques do not take traffic load dynamics into account, and thus, when applied to traffic load estimation, they cannot guarantee that the sampling error in each block falls within a prescribed

---

[3] Though systematic sampling technique does not give any assessment of error, when the packet sizes are random in the arriving order, it simulates random sampling.

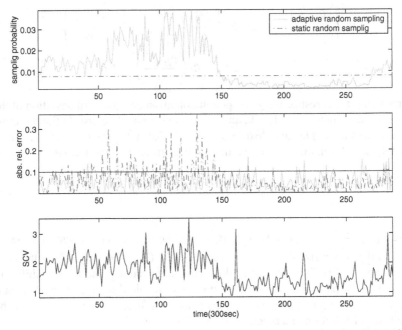

**Fig. 2.4** Importance of adaptiveness: adaptive vs. static random sampling ($\{\eta, \varepsilon\} = \{0.1, 0.1\}$).

error tolerance level. Furthermore, it is difficult to determine what is the appropriate fixed sampling probability (or the value for N in "one-out-of-N" sampling) to be used for all blocks.

To help illustrate the importance of adjusting sampling probability to packet size variability, in Figure 2.4 we compare the *optimal* adaptive random sampling technique to the static random sampling technique using the Auckland trace $\Pi_1$ shown in Table 2.2. To make a fair comparison, the fixed sampling probability for the static random sampling technique is set such that the *sampling fraction* (i.e., the amount of sampled data) over the entire trace is the same as that under the optimal adaptive random sampling technique. The top plot in Figure 2.4 shows the optimal sampling probability used by the adaptive sampling technique over time (the block size $B = 300sec$) as well as the fixed sampling probability used by the static random sampling. The middle plot shows the resulting relative errors by both sampling techniques. The bottom plot shows the *SCV* of the packet sizes across the blocks.

From the figure we see that when the variability of packet size distribution of a block is large, static random sampling tends to *undersample* packets, resulting in large estimation errors. This may lead to false alarm or non-detection by a load change detection algorithm. On the other hand, when the variability of packet size distribution of a block is small, static random sampling tends to *oversample* packets, thereby wasting processing capacity and memory space of the measurement device. Moreover, the frequent oscillation between oversampling and undersampling of static random sampling causes an undesirable increase in the variance of estimation errors. This example demonstrates that in order to ensure a desired accuracy in traffic load estimation without resorting to unnecessary oversampling, packet sampling probability for each block must be adjusted in accordance with the traffic load dynamics. This is the essential idea behind our proposed adaptive random sampling technique. The key challenge which remains to be addressed is how to determine the (optimal) sampling probability for each block *without a priori knowledge* of the traffic parameters – the *SCV* of packet size distribution and packet count of a block. The next section is devoted to the analysis and solution of this problem.

## 2.4 Adaptive Random Sampling with Bounded Errors

In this section, we present an AR (Auto-Regressive) model for predicting two key traffic parameters for traffic load estimation – the *SCV* of packet size distribution and packet count of a block – using past (sampled) data from previous blocks. The AR model is justified by empirical studies using real network traffic traces. In addition to estimation errors due to sampling, the prediction model also introduces *prediction* errors. We quantify and analyze the impact of these errors on the traffic load estimation and discuss how these errors can be controlled.

### 2.4.1 AR Model for Traffic Parameter Prediction

**Table 2.1** Notation.

| Notation | Explanation |
|---|---|
| $S_k$ | *SCV* of the population of $k$th block |
| $S_k^s$ | *SCV* of the samples of $k$th block |
| $\hat{S}_k^s$ | predicted *SCV* of the samples of $k$th block |
| $n_k^*$ | minimum number of samples needed in $k$th block |
| $\hat{n}_k$ | predicted minimum number of samples needed in $k$th block |
| $n_k$ | actual number of samples in $k$th block |
| $m_k$ | actual number of packets in $k$th block |
| $\hat{m}_k$ | predicted number of packets in $k$th block |

The efficacy of prediction depends on the correlation between the past and future values of the parameters being predicted. We have analyzed many public-domain real network traffic traces, and a subset of traces we studied is listed in Table 2.2. We found that the *SCV*'s of the packet sizes of two *consecutive* blocks are strongly correlated; the same is also true for the packet counts, $m$'s, of two *consecutive* blocks. As an illustration, Figures 2.5(a) and 2.5(b) show, respectively, the scatter plots of *SCV* and $m$ of two consecutive blocks (the block size $B = 60sec$) using the trace $\Pi_4$

in Table 2.2. It is evident that the values of *SCV* and *m* of two consecutive blocks are highly correlated, in that the correlation coefficients are very close to 1. In fact, there is a strong *linear* relationship between these values.

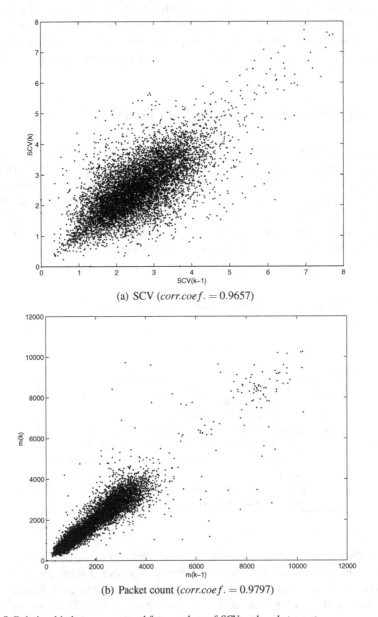

(a) SCV (*corr.coef.* = 0.9657)

(b) Packet count (*corr.coef.* = 0.9797)

**Fig. 2.5** Relationship between past and future values of *SCV* and packet count.

The predictability may depend on the time scale (block size) of observation. We observed strong positive correlations for a wide range of time scale from $1min$ to $30min$ for long traces and $1sec$ to $10sec$ for short ($90sec$) traces. As a further justification, we remark that the predictability of network traffic has also been studied by other researchers. For instance, in [19] the authors investigated the questions of how far into the future a traffic rate process can be predicted for a given error constraint, and how much the prediction error is over a specified number of future time intervals (or steps). They showed that prediction works well for one step into the future, although the prediction accuracy degrades quickly as the number of steps increases. In the context of our work, note that we only need to predict the traffic parameters for the next step (i.e., the next block).

The strong linear relationship evident in Figures 2.5(a) and 2.5(b) suggests that linear regression can be used for the prediction of the $SCV$ of packet sizes and packet count $m$ of a future block using the values of the previous blocks. We employ an AR (Auto-Regressive) model for predicting the traffic parameters $SCV$ and $m$, as compared to other time series models, and the AR model is easier to understand and computationally more efficient. In particular, using the AR model, the model parameters can be obtained by solving a set of simple linear equations [12], making it suitable for online traffic load estimation. In the following we formally describe the AR model for the traffic parameter prediction.

We first present an AR($u$) model for predicting the $SCV$ of the next block using the $SCV$ of *sampled* packet sizes of the $u$ previous blocks. The notation used here and in the rest of this chapter is summarized in Table 2.1. Let $S_k$ be the $SCV$ of the packet sizes in the $k$th block, and $S_k^s$ be the $SCV$ of the *packet sizes randomly sampled* in the $k$th block. We can relate $S_k$ and $S_k^s$ as follows:

$$S_k^s = S_k + Z_k \tag{2.6}$$

where $Z_k$ denotes the error in estimating the actual $SCV$ of the packet sizes using the random packet samples. (We refer to $Z_k$ as the estimation error.)

Using the AR($u$) model [12], $S_k^s$ can be expressed as

$$S_k^s = \sum_{i=1}^{u} a_i^s S_{k-i}^s + e_k^s \tag{2.7}$$

where $a_i$, $i = 1, \ldots, u$, are the model parameters, and $e_k^s$ is the *uncorrelated* error (which we refer to as the *prediction error*). The error term $e_k^s$ follows a normal distribution with mean 0 and variance $var(e_k^s) = \sigma_{S_k^s}^2 (1 - \sum_{i=1}^{u} a_i^s \rho_{S_k^s, i})$. Here $\rho_{S_k^s, i}$ is the lag-$i$ autocorrelation of $S_k^s$'s. The model parameters $a_i$, $i = 1, \ldots, u$, can be determined by solving a set of linear equations (2.8) in terms of $v$ past values of $S_i^s$'s, where $v \geq 1$ is a configurable parameter independent of $u$, and is typically referred to as the memory size.

$$\rho_h = \sum_{i=1}^{u} a_i \rho_{h-i}, \text{ where } h = v, \ldots, v - u + 1 \text{ and } \rho_h \text{ is lag-}h \text{ autocorrelation of the data}$$

$$(2.8)$$

Using the above AR($u$) model, at the end of the $(k-1)$th block, we predict the *SCV* of the $k$th block using the *SCV* values of the sampled packet sizes of the $u$ previous blocks as follows:

$$\hat{S}_k^s = \sum_{i=1}^{u} a_i^s S_{k-i}^s. \qquad (2.9)$$

Combining (2.6), (2.7) and (2.9), we have

$$\hat{S}_k^s = S_k + Z_k + e_k^s. \qquad (2.10)$$

Hence we see that there are two types of errors in predicting the actual *SCV* of the packet size of the next block using the sampled packet sizes of the previous blocks: the estimation error $Z_k$ due to random sampling, and the prediction $e_k^s$ introduced by the prediction model. The total resulting error is $Z_k + e_k^s$. In Section 2.4.2 we analyze the properties of these errors and their impact on the traffic load estimation.

We now briefly describe how the packet count $m_k$ of the $k$th block can be estimated based on the past packet counts using the AR($u$) model. Let $m_k$ denote the packet count of the $k$th block, then by using the AR($u$) model, we have $m_k = \sum_{i=1}^{u} b_i m_{k-i} + e_{m,k}$, where $b_i$, $i = 1, 2, \ldots, u$, are the model parameters as before, and $e_{m,k}$ is the prediction error term, which is normally distributed with zero mean. Let $\hat{m}_k$ denote the *predicted* packet count of the $k$th block. Using the the AR($u$) prediction model, we have $\hat{m}_k = \sum_{i=1}^{u} b_i \hat{m}_{k-i}$.

As in the case of predicting *SCV* of the packet sizes using the AR($u$) prediction model, the prediction of the packet count using the past *sampled* packet counts introduces both estimation error and prediction error. However, in the case of predicting the packet count $m$, it is *not* unreasonable to assume that the *actual* packet count of a block is known at the end of the block. This is because in the modern commercial router design, a packet counter is often included in the line card of a router, as such a packet counter does not overly burden a router in terms of both processing and memory capacities[4].

In this case, we can predict the packet count of the next block using the *actual* packet counts of the previous blocks. Namely, $\hat{m}_k = \sum_{i=1}^{u} b_i m_{k-i}$. Hence, only the prediction error is involved when a packet counter is available. For simplicity, we will assume that this is the case in our work. Note that this assumption does not change the nature of the adaptive random sampling technique we proposed, but only simplifies the analysis of the sampling errors.

---

[4] Observe that the packet count of a block can be collected without inspecting the contents of a packet. Hence it does not cause a significant burden on the routers. For example, consider a link with bandwidth $10Gbps$. Suppose the worst case is where only the smallest IP packets (40 bytes) have arrived. Then, there can be at most $1.875G$ packets in a block of 60 seconds. The size of the counter needed is only 32 bits. If we assume that $I$ instructions are needed to increment the counter, then we need only $31.25 * I$ MIPS for maintaining the packet counter.

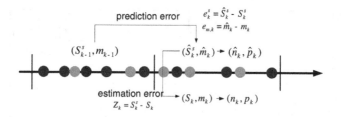

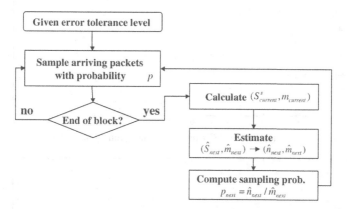

**Fig. 2.6** Traffic parameter prediction process.

**Fig. 2.7** Flow chart of adaptive random sampling.

Given the predicted *SCV* of the packet size distribution and packet count of the next block, we can now calculate the (predicted) minimum number of required packet samples using (2.4) and the sampling probability for the next block:

$$\hat{n}_k = z_p \hat{S}_k^s \text{ and } \hat{p}_k = \frac{\hat{n}_k}{\hat{m}_k}. \tag{2.11}$$

The entire process of predicting traffic parameters *SCV* and *m* is depicted in Figure 2.6. Figure 2.7 shows the flow chart of the adaptive random sampling procedure. Using the AR prediction model, at the end of each block, the model parameters ($a_i$'s for *SCV*, $b_i$'s for *m*) need to be computed [12]. The complexity of the AR prediction model parameter computation is only $O(v)$ where $v$ is the memory size. Through empirical studies, we have found that small values of the memory size (around 5) are sufficient to yield good prediction.

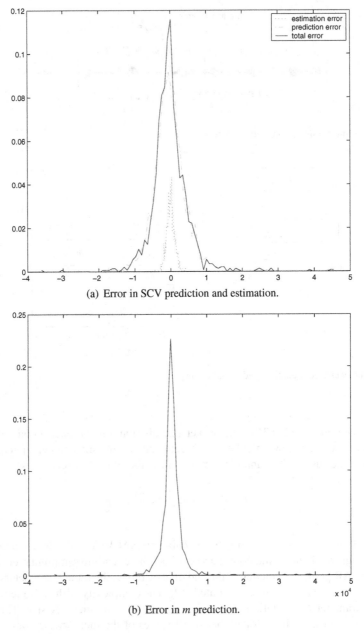

(a) Error in SCV prediction and estimation.

(b) Error in *m* prediction.

**Fig. 2.8** Gaussian prediction error.

## 2.4.2 Analysis of Errors in Traffic Load Estimation via Sampling

In this subsection we analyze the impact of estimation and prediction errors on the traffic load estimation. We first study the properties of the errors introduced by the adaptive random sampling process. We then establish several lemmas and theorems to quantify the impact of these errors on the relative error in the traffic load estimation.

Recall from (2.10) that there are two types of errors in estimating the *SCV* of the packets size of the next block using the past sampled packet sizes: the estimation error $Z_k$ and the prediction error $e_k^s$. From empirical studies using real network traces, we have found that the errors generally follow a normal distribution with mean 0. An example using the trace $\Pi_1$ is shown in Figure 2.8(a), where we see that both the estimation error and prediction error as well as the total error $(Z_k + e_k^s)$ have a Bell-shape centered at 0. We have performed the skewness test and kurtosis test [2], and these tests conform the normality of these errors. Similar empirical studies have also shown that the error $(e_{m,k})$ in the packet count prediction is also normally distributed with zero mean. See Figure 2.8(b) for an example using the same network traffic trace as in Figure 2.8(a).

The above results suggest that we can approximate both the estimation error and prediction error using normal distributions with zero mean. This allows us to quantify the variance of the errors introduced by the adaptive random sampling process. For example, assume, for simplicity, that an AR(1) model is used for predicting $S_k$, the *SCV* of the packet sizes of the $k$th block. Then the variance of the prediction error, $var(e_k^s)$, is given by $var(e_k^s) = \sigma_{S_k^s}^2(1 - a_1^s \rho_{S_k^s,1})$, where $\rho_{S_k^s,1}$ is the lag-1 autocorrelation of $S_k^s$. From (2.6) and (2.7), we have $var(Z_k) = (a_1^s)^2 var(S_{k-1}^s) + var(e_k^s) - var(S_k)$. Given sufficient packet samples, $var(S_{k-1}^s) \approx var(S_k)$. Thus, $var(Z_k) = \sigma_{S_k^s}^2((a_1^s)^2 - a_1^s \rho_{S_k^s,1})$. Therefore, the variance of the total error in predicting $S_k$ is

$$var(Z_k) + var(e_k^s) = \sigma_{S_k^s}^2(1 - 2a_1^s \rho_{S_k^s,1} + (a_1^s)^2). \qquad (2.12)$$

We now quantify the impact of these errors on the relative error in the traffic load estimation. Define $n_k = m_k \cdot \frac{\hat{n}_k}{\hat{m}_k}$, which is the *actual* number of packets randomly sampled (on the average) in the $k$th block, given the (predicted) minimum sampling probability $\hat{p}_k = \hat{n}_k / \hat{m}_k$. Then the estimated traffic load of the $k$th block is given

$$\tilde{V}_k = \frac{m_k}{n_k} \sum_{j=1}^{n_k} \hat{X}_j, \qquad (2.13)$$

where $\hat{X}_j$ denotes the packet size of the $j$th randomly sampled packet in the $k$th block.

Using the central limit theorem for a sum of a random number of random variables (see p.369, problem 27.14 in [4]), we can establish the following two lemma and theorem.

**Lemma 2.1.** $\frac{n_k}{n_k^*}$ converges to *1 almost surely as* $n_k^* \to \infty$.

*Proof.* For simplicity of notation, we drop the subscript $k$ in the notation. Note first that

$$\lim_{m \to \infty} \frac{m}{\hat{m}} = \lim_{m \to \infty} \frac{m}{m + e_m} = 1.$$

Since $S^s \to S$ as $n \to \infty$, $\hat{n} = z_p S^s \to z_p S = n$.

From $\tilde{n}_k = \hat{n} \cdot \frac{m}{\hat{m}}$, we have $\lim_{n \to \infty} \frac{\tilde{n}}{n} = 1$. ∎

**Lemma 2.2.** *With probability* $1 - \eta$, *the relative error in estimating the traffic load* $V_k$ *of the kth block is*

$$\left| \frac{\tilde{V}_k - V_k}{V_k} \right| \leq \varepsilon + \frac{1}{\sqrt{z_p}}(1+\varepsilon)Y + o\left(\frac{1}{m}\right)$$

$$\approx \varepsilon + \frac{1}{\sqrt{z_p}}(1+\varepsilon)Y$$

*where recall that* $z_p = \left( \frac{\Phi^{-1}(1-\eta/2)}{\varepsilon} \right)^2$, *and Y is a normally distributed random variable with mean 0 and variance 1, i.e.,* $Y \sim N(0,1)$.

*Proof.*

$$\tilde{V} = \frac{m}{\tilde{n}} \sum_{i=1}^{\tilde{n}} X_i$$

$$= \frac{m}{\tilde{n}} \left( \tilde{n}\mu^s + \sigma^s \sqrt{n}Y + o(\sqrt{n}) \right)$$

$$= \hat{V} + \frac{(\sigma^s \sqrt{n}Y + o(\sqrt{n}))\hat{m}}{\hat{n}}$$

$$\approx \hat{V} + \frac{\hat{m}}{z_p \hat{S}^s} \sqrt{S^s} \mu^s \sqrt{z_p S} Y$$

$$= \hat{V} + \frac{\hat{m}\mu^s}{\sqrt{z_p}} \cdot \frac{\sqrt{S(S+Z)}}{S+Z+e^s} Y$$

where, $Z$ is a normal random variable with mean 0, and $Y$ is standard normal random variable.

Note that

$$\frac{\sqrt{S(S+Z)}}{S+Z+e^s} \leq \frac{S+Z/2}{S+Z+e^s}$$

$$\leq \frac{S+Z/2}{S+Z} \leq 1$$

Since $\left| \frac{\hat{V}-V}{V} \right| < \varepsilon$ with probability $1 - \eta$, the following holds with probability $1 - \eta$.

$$\left| \tilde{V} - V \right| < V\varepsilon + \frac{V}{\sqrt{z_p}}(1+\varepsilon)Y + \frac{e_m \mu^s}{\sqrt{z_p}}Y \qquad (2.14)$$

Therefore, the relative error is given by

$$\left|\frac{\tilde{V}-V}{V}\right| < \varepsilon + \frac{1}{\sqrt{z_p}}(1+\varepsilon)Y + \frac{e_m\mu^s}{V\sqrt{z_p}}Y$$

$$= \varepsilon + \frac{1}{\sqrt{z_p}}(1+\varepsilon)Y + o(\frac{1}{m})$$

$$\approx \varepsilon + \frac{1}{\sqrt{z_p}}(1+\varepsilon)Y$$

■

Lemma 2.2 yields a theoretic bound on the variance of adaptive random sampling, i.e.,

$$var\left(\left|\frac{\tilde{V}-V}{V}\right|\right) \le \frac{(1+\varepsilon)^2}{z_p} \text{ with probability } 1-\eta. \tag{2.15}$$

Notice that the variance of adaptive random sampling is independent of the distribution of objects being sampled and is *controllable* by the accuracy parameter. On the other hand, the variance of static random sampling depends on the *SCV* and the number of samples. i.e.,

$$var\left(\frac{\hat{V}-V}{V}\right) = var\left(\frac{\frac{m}{n}\sum_{i=1}^{n}\hat{X}_i - \sum_{j=1}^{m}X_j}{\sum_{j=1}^{m}X_j}\right) = var\left(\frac{\sum_{i=1}^{n}\hat{X}_i}{n\mu}\right)$$

$$= \left(\frac{1}{n\mu}\right)^2 \cdot n \cdot \sigma^2 = \frac{\sigma^2}{\mu^2}\cdot\frac{1}{n} = \frac{S}{n} \tag{2.16}$$

The variance bound (2.15) of adaptive random sampling suggests that in order to accommodate the prediction and estimation errors introduced by the traffic parameter predictions, we can replace the error bound $\varepsilon$ by a tighter bound $\varepsilon'$:

$$\varepsilon' = \varepsilon - s \cdot \frac{(1+\varepsilon)}{\sqrt{z_p}} \tag{2.17}$$

where $s$ is a small adjustment parameter that can be used to control the variance of the relative error.

## 2.5 Experimental Results

In this section we empirically evaluate the performance of our adaptive random sampling technique using the real network traces. The traces used in this study are obtained from CAIDA [5], and their statistics are listed in Table 2.2. In this study we have primarily used the *long* duration traces (the Auckland-II traces) that give enough number of estimates to produce more sound statistics and reliable results. But we have also investigated the short duration traces from the higher speed links in the backbone. We believe that the efficacy of our adaptive random sampling technique as demonstrated in this section are applicable to other traces. For consistency of illustration, the results shown in this section are based on the trace $\Pi_1$ unless otherwise specified.

**Table 2.2** Summary of traces used.

| Trace name | Trace | Arrival rate | Duration |
|---|---|---|---|
| $\Pi_1$ | Auckland-II 19991201-192548-0 | 92.49KBps | 24h 02m 58sec |
| $\Pi_2$ | Auckland-II 19991201-192548-1 | 55.16KBps | 24h 02m 57sec |
| $\Pi_3$ | Auckland-II 19991209-151701-1 | 49KBps | 23h 11m 38sec |
| $\Pi_4$ | Auckland-II 20000117-095016-0 | 168KBps | 2h 23m 15sec |
| $\Pi_5$ | Auckland-II 20000114-125102-0 | 222.14KBps | 21m 37sec |
| $\Pi_6$ | AIX (OC12c) 989950026-1 | 25.36MBps | 90sec |
| $\Pi_7$ | AIX (OC12c) 20010801-996689287-1 | 21.60MBps | 90sec |
| $\Pi_8$ | COS (OC3c) 983398787-1 | 4.95MBps | 90sec |

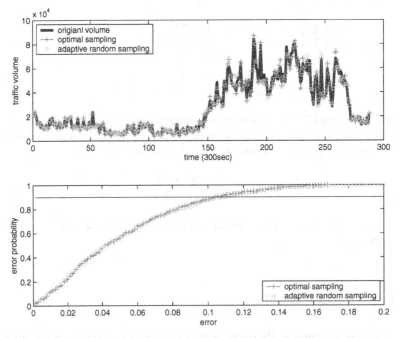

**Fig. 2.9** Adaptive random sampling: traffic volume estimation and bounded relative error ($\{\eta,\varepsilon\} = \{0.1, 0.1\}, B = 300sec$).

To show the effectiveness of the prediction model used in our adaptive random sampling technique, we first compare the performance of our technique with that of the *ideal* optimal sampling. In the ideal optimal sampling, the optimal sampling probability for each block is computed using (2.5), assuming that the *SCV* of the packet sizes and packet count of the block are known. The results are shown in Figure 2.9. The figure on the top shows the time series of the original traffic load, the estimated traffic loads using both the ideal optimal sampling and the adaptive random sampling with prediction. For the accuracy parameters of $\{\eta, \varepsilon\} = \{0.1, 0.1\}$, the series are very close and hardly differentiable visually. The figure on the bottom shows the cumulative probability of relative errors in traffic load estimation for both the ideal optimal sampling and adaptive random sampling with prediction. The horizontal line in the figure indicates the $(1 - \eta)$th quantile of the errors. We see that for both of the sampling methods, the traffic load estimation indeed conforms to the pre-specified accuracy parameter, i.e., the probability of relative errors larger than $\varepsilon = 0.1$ is around $\eta = 0.1$.

To further investigate the performance of the adaptive random sampling with prediction, in Figure 2.10(a) we vary the error bound $\varepsilon$ (while fixing $\eta$ at 0.1), and plot the corresponding $(1 - \eta)$th quantile of relative errors. We see that the $(1 - \eta)$th quantiles of relative errors for the whole range of the error bound $\varepsilon$ stay close to the prescribed error bound. For comparison, in the figure we also plot the corresponding results obtained using the static random sampling. To provide fair comparison, the (fixed) sampling probability of the static random sampling is chosen such that the *sampling fraction* (or the total amount of sampled data) over the entire trace is the same as that of the adaptive random sampling. We see that for all ranges of the error bound, the static random sampling produces a much larger the $(1 - \eta)$th quantile of relative errors.

Another key metric for comparing sampling techniques is the variance of an estimator [17]. Small variance in estimation is a desired feature of a sampling method in that the estimate is more reliable when used in place of the value of a population. This feature is especially important when the sampling method is applied to change point detection, since large variation in estimation may cause outliers in the estimated signal, making it difficult to detect change points (see discussion in Section 2.6). In Figure 2.10(b), we compare the standard deviation of the relative errors in traffic load estimation for both the adaptive random sampling and the static random sampling. As the figure shows, the variation of errors of the adaptive random sampling is always bounded within the theoretic upper bound (2.15). On the contrary, due to frequent excessive undersampling and oversampling (as noted in Section 2.3), the static random sampling has a much larger variation of errors. In particular, the error variance of the static random sampling is always larger than the theoretic variance bound for the adaptive random sampling. In summary, the above results demonstrate the superior performance of our adaptive random sampling technique over the static random sampling. We now compare the adaptive random sampling and static random sampling in terms of their *resource efficiency*. We measure the resource efficiency using the sampling fraction – the ratio of the total amount of sampled data produced by a sampling technique over the total amount data in a trace. Sampling fraction provides an indirect measure of the processing and storage requirement of a sampling technique. To compare the adaptive random

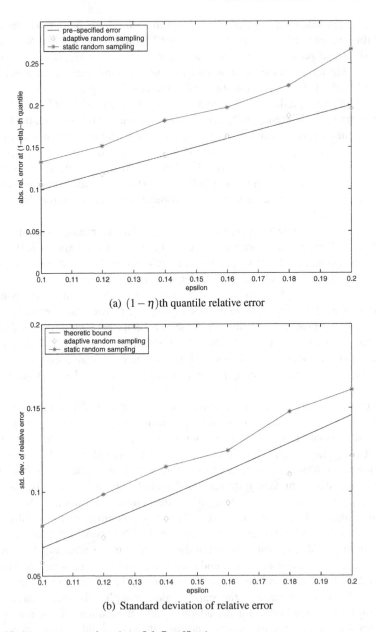

(a) $(1 - \eta)$th quantile relative error

(b) Standard deviation of relative error

**Fig. 2.10** Accuracy comparison ($\eta = 0.1, B = 60sec$).

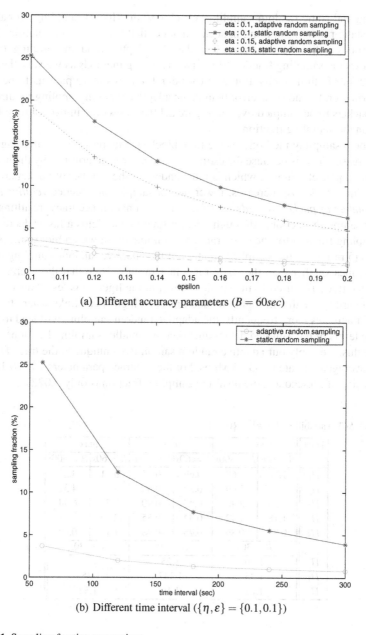

(a) Different accuracy parameters ($B = 60sec$)

(b) Different time interval ($\{\eta, \varepsilon\} = \{0.1, 0.1\}$)

**Fig. 2.11** Sampling fraction comparison.

sampling and static random sampling, we choose the (fixed) sampling probability for the static random sampling in such a manner that the $(1 - \eta)$th quantile of relative errors satisfies the same error bound as the adaptive random sampling. Figure 2.11 shows the sampling fraction of the two sampling methods as we vary the error bound $\varepsilon$. For both methods, tighter error bound requires more packets to be sampled. However, for the same error bound, the adaptive random sampling requires far fewer packets to be sampled overall. Figure 2.11(b) shows the impact of time block size $B$ on the sampling fraction.

For both sampling methods, as the time block size increases, the sampling fraction decreases. This is because the estimation accuracy is determined by the number of required packet samples, which is independent of the number of packet arrivals. As the time block size increases, fewer packet samples are needed *relative to the total number of packet arrivals* to achieve the estimation accuracy, resulting in a smaller sampling fraction. Although a larger time block yields a faster decrease in the sampling fraction for the static random sampling, even with a block size of 300 seconds (5 minutes), the sampling fraction of the adaptive random sampling is still several times smaller than the static random sampling. Note that the average data rate of the trace $\Pi_1$ (used in the studies shown in the figures) is less than 1 MBps. It is not hard to see that in highly loaded links and high speed links where the traffic load fluctuates more frequently, the adaptive random sampling will lead to more gains in terms of the sampled data reduction (i.e., smaller sampling fraction). To illustrate this, we apply our adaptive random sampling technique to the trace $\Pi_6$ that has an average data rate of 35.36 Mpbs. For the accuracy parameters $\{0.1, 0.1\}$ and a block size of 30 seconds, the resulting sampling fraction is only 0.022%!

**Table 2.3** *SCV* variability ($\{\eta, \varepsilon\} = \{0.1, 0.1\}$).

| Trace | avg. | | min. | | max. | |
|---|---|---|---|---|---|---|
| | $B = 60s$ | $B = 300s$ | $B = 60s$ | $B = 300s$ | $B = 60s$ | $B = 300s$ |
| $\Pi_1$ | 1.70 | 1.68 | 0.69 | 0.85 | 5.21 | 3.52 |
| $\Pi_2$ | 2.11 | 2.09 | 0.39 | 0.60 | 5.29 | 4.32 |
| $\Pi_3$ | 2.59 | 2.48 | 0.47 | 0.90 | 7.39 | 6.44 |
| $\Pi_4$ | 1.18 | 1.12 | 0.32 | 0.55 | 2.55 | 1.38 |
| $\Pi_5$ | 0.89 | 0.89 | 0.75 | 0.78 | 1.13 | 0.99 |
| | $B = 30s$ | | $B = 30s$ | | $B = 30s$ | |
| $\Pi_6$ | 1.26 | | 1.25 | | 1.27 | |
| $\Pi_7$ | 1.35 | | 1.347 | | 1.353 | |
| $\Pi_8$ | 1.22 | | 1.16 | | 1.32 | |

To conclude this section, we provide a more detailed study of the *SCV* of packet sizes in the real network traffic traces. Understanding the *SCV* of packet size distribution is important, as the number of required packet sizes is proportional to *SCV*. Thus, the *SCV* of packet sizes in a network traffic trace has a direct impact on the resulting sampling fraction. The *SCV* statistics of the traces are presented in Table 2.3. We see that the *SCV*s of packet sizes of the traces vary significantly, although some

**Table 2.4** Sampling fraction ($\{\eta,\varepsilon\} = \{0.1,0.1\}$).

| Trace | Sampling fraction (%) | |
|---|---|---|
| | $B = 60s$ | $B = 300s$ |
| $\Pi_1$ | 5.48 | 0.67 |
| $\Pi_2$ | 5.85 | 0.83 |
| $\Pi_3$ | 5.07 | 1.01 |
| $\Pi_4$ | 1.67 | 0.30 |
| $\Pi_5$ | 0.91 | 0.25 |
| | $B = 30s$ | |
| $\Pi_6$ | 0.022 | |
| $\Pi_7$ | 0.022 | |
| $\Pi_8$ | 0.12 | |

of the traces ($\Pi_1$ - $\Pi_5$) are captured over the same physical link over different time. For the accuracy parameters of $\{\eta,\varepsilon\} = \{0.1,0.1\}$, the sampling fractions for these traces (with block size $B = 60sec$ or $300sec$) are also listed in Table 2.4. It is clear that the SCV of packet sizes of the traces has a direct impact on the sampling fraction. In general, a smaller SCV leads to a smaller sampling fraction. Furthermore, the data arrival rate of the traces also affects the sampling fraction. For example, the traces $\Pi_6$ and $\Pi_8$ have similar SCV's. However, the average data rate of $\Pi_6$ is about five times faster than that of $\Pi_8$ (see Table 2.2). As a result, the sampling fraction of $\Pi_6$ is about 5 times smaller than that of $\Pi_8$.

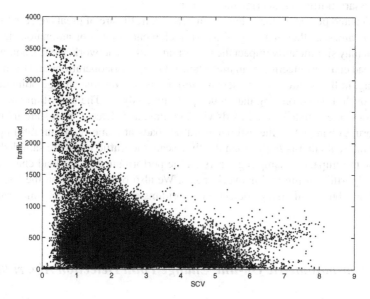

**Fig. 2.12** Traffic load vs. SCV.

Finally, Figure 2.12 shows the relation between *SCV* and traffic load in a scatter plot using the traces. We observe that when the link is more highly utilized (i.e., larger traffic load), the *SCV* of packet sizes tends to be smaller. Since the lower *SCV* leads to a smaller number of required packet samples, the adaptive random sampling is likely to offer a higher rate of sampled data reduction in times of high load, while providing the desired degree of accuracy in traffic load estimation. In other words, when a high-speed link is highly utilized, our adaptive sampling technique results in fewer packets to be sampled, thereby reducing the burden (both in terms of processing and storage) on the traffic monitoring and measurement device (whether on-board a router or off-board). And this is achieved *without sacrificing the sampling accuracy*! Lastly, we would like to point out that although our adaptive random sampling technique is designed with the application to load change detection in mind, it can also be applied to other traffic engineering applications.

## 2.6 Load Change Detection with Sampled Measurement

Many practical problems arising in network performance monitoring and management are due to the fact that changes in network conditions are observed too late. Moreover, the exact time of a change may not be readily available. Such information on the change point in traffic load can help locate the source of change and initiate an appropriate action to deal with the change. In other words, detection of abrupt changes is an important first step towards reacting to changes by invoking the necessary traffic engineering mechanisms.

Sudden and persistent load changes in network traffic are of great concern to network operators, as they may signal network element failures or anomalous behaviors, and may significantly impact the performance of the network. Hence, automatic traffic load change detection is an important aid in network operations and traffic engineering. In this section, we present a *non-parametric, on-line* change point detection algorithm based on singular value spectrum analysis. The algorithm takes the time series of estimated traffic loads via sampling, and detects "non-stationarities" (i.e., abrupt changes) in the estimated traffic loads at some (configurable) operational time scale that is *larger* than the time scale the traffic loads are sampled. We examine the impact of sampling errors on the performance of the load change detection algorithm using real network traces. We also briefly touch on the issues in designing robust load change detection algorithms and the impact of time scale of change.

### 2.6.1 Non-parametric On-line Change Point Detection Algorithm

In the problem of traffic load change detection, we assume that the statistics of traffic loads are normally either constant or slowly time-varying; otherwise an *abrupt*

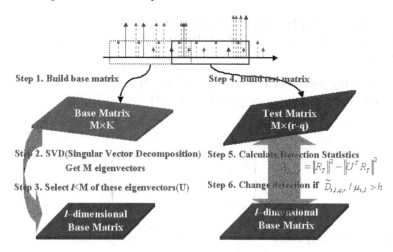

**Fig. 2.13** Change detection algorithm.

*change* should be recognized. By abrupt changes, we mean changes in characteristics that occur very fast, if not instantly. But before and after the change, the properties are fairly stationary with respect to the time scale of interest. Note that abrupt changes by no means imply changes with large magnitude. Many network management concerns are related with detection of small changes. Traditional *parametric* techniques involve estimation of certain parameters of the time series such as mean and variance and some presumed distributions (e.g., Gaussian) of the parameters for assessing statistical significance of these estimates. However, such assumptions typically cannot be applied to real network traffic [22]. In particular, a few short bursts (outliers) may greatly distort the estimates. Thus, traditional parametric techniques may not work well in the presence of outliers. A non-parametric algorithm based on singular-spectrum analysis is much more robust, since it efficiently separates noise (outliers) from the signal. Since sampling may further introduce or magnify outliers, tolerance of noise is critical in our framework. In addition, amenability to on-line implementation is another consideration in selecting traffic load change detection algorithms. In our study we employ such a non-parametric change point detection algorithm based on singular spectrum analysis (SSA). The algorithm is developed in [16], which we briefly describe below.

Let $y_1, y_2, \ldots$ be a time series of estimated traffic loads. (Note that the time index $t$ here is in the unit of time block $B$ of sampling.) For each time $t = 1, 2, \ldots$, part of the time series, $y_{t+1}, \ldots, y_{t+N}$, is considered as the *base data*. Another sub-series, $y_{t+q+1}, \ldots, y_{t+r+M-1}$, where $q > 0$, is called the "test data." Intuitively, if there is no significant change in the essential traffic signals, the *distance* in statistics between the test data and the base data should stay reasonably small. If the distance in statistics is larger, it signals an abrupt change. The basic idea of the change point detection algorithm is depicted in Figure 2.13. The steps involved in the change point detection procedure are given below (please refer to [16] for more details).

Let $N, M, l, p$ and $q$ be some integers such that $M \leq N/2$, $0 \leq q < r$ and $N \leq r$. An integer $K$ is set to be $K = N - M + 1$. For each time index $t = 0, 1, \ldots$, compute the following:

1. Build the lag-covariance matrix $R_B{}^{(t)} = \frac{1}{K} Y_B{}^{(t)} (Y_B{}^{(t)})^T$ of the trajectory matrix $Y^{(t)}$ with the base data (as shown in Figure 2.14).

$$Y_B{}^{(t)} = \begin{pmatrix} y_{t+1} & y_{t+2} & \cdots & y_{t+K} \\ y_{t+2} & y_{t+3} & \cdots & v_{t+K+1} \\ \vdots & \vdots & \ddots & \vdots \\ y_{t+M} & y_{t+M+1} & \cdots & y_{t+N} \end{pmatrix}$$

**Fig. 2.14** Base trajectory matrix.

$$Y_T{}^{(t)} = \begin{pmatrix} y_{t+q+1} & y_{t+q+2} & \cdots & y_{t+r} \\ y_{t+q+2} & y_{t+q+3} & \cdots & y_{t+r+1} \\ \vdots & \vdots & \ddots & \vdots \\ y_{t+q+M} & y_{t+q+M+1} & \cdots & y_{t+r+M-1} \end{pmatrix}$$

**Fig. 2.15** Test trajectory matrix.

2. Perform SVD (Singular Value Decomposition) of $R_B^{(t)}$: $R_B^{(t)} = U \Lambda U^T$.
3. Determine a $l$-dimensional subspace spanned by the first $l$ eigenvectors ($U_l$) of $R_B^{(t)}$.
4. Similarly, build the lag-covariance matrix $R_T{}^{(t)} = \frac{1}{K} Y_T{}^{(t)} (Y_T{}^{(t)})^T$ of the trajectory matrix with the training data.
5. Compute the detection statistics $\mathscr{D}_{t,l,q,r}$, the sum of the squared Euclidean distances between the vectors $Y_j^{(t)}$ ($j = q+1, \ldots, r$) and $U_l$. i.e., $\|R_T\|^2 - \|U_l^T R_T\|^2$.
6. Decide if there is an abrupt change ($\mathscr{D}_{t,l,q,r} > threshold$), and generate an alarm with estimated time $\tau(= t + r + M - 1)$ of change with the detection statistic $\mathscr{D}_{t,l,q,r}$.

## 2.6.2 Experiments

The detection statistics $\mathscr{D}_{t,l,q,r}$ holds *asymptotic normality* under the conditions that the window size $N$ and the lag $M$ are sufficiently large [16]; from this result, an

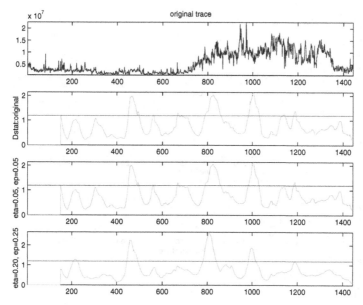

**Fig. 2.16** Detection statistics for population and estimated traffic load.

asymptotic probability of a change can also be derived. In the results shown in this section, we use a 99% of significance level as the detection threshold. Figure 2.16 illustrates the impact of sampling errors on the performance of the load change detection algorithm. The second plot of Figure 2.16 depicts the detection statistics of the time series of the original traffic loads (with a time block of $B = 60sec$), which is shown on the top row. The third and fourth plots are the detection statistics of the estimated traffic loads with the sampling accuracy parameters $\{\eta = 0.05, \varepsilon = 0.05\}$ and $\{\eta = 0.20, \varepsilon = 0.25\}$, respectively. With the sampling accuracy parameters $\{\eta = 0.05, \varepsilon = 0.05\}$, all of the changes detected using the original traffic loads are also detected using the estimated traffic loads. However, with the sampling accuracy parameters $\{\eta = 0.20, \varepsilon = 0.25\}$, one false-alarm (around 200*min*) is generated, and two small load changes in the neighborhood of 600*min* and another small load change around 1200*min* are not detected. This evidently tells us that *controlling estimation errors* are critical in traffic load change detection. Note here that the detection algorithm also finds subtle load change points that would otherwise be undetectable via visual inspection by humans without further data processing.

The parameters in the detection algorithm can be tuned to control the time scale of load changes as well as sensitivity (or magnitude) of load changes that are of interest to network operators. Depending on the type of load changes we are looking for, the window size of the base and test matrices, $M$, and the location and length of the test data $(q, r)$ can be configured accordingly. Observe that if $M$ is too small, then an outlier may be recognized as a structural change. Hence, it is recommended that $M$ is chosen to be sufficiently large but smaller than the time scale of load changes to

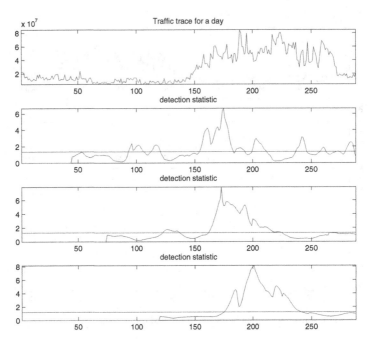

**Fig. 2.17** Detection statistics with varying lag parameter.

be detected, so as not to miss out on all the changes in the time series of traffic loads. In Figure 2.17, we show the effect of varying $M$ on the detection statistics. For ease of observation, we use a larger sampling time block ($B = 300sec$). The top plot in Figure 2.17 shows the time series of the original traffic loads. The detection statistics corresponding to $M = 30, 50, 80$ are shown, respectively, in the second, third, and fourth plot. We see that using smaller $M$ detects load changes that occur in a smaller time scale. For example, close visual inspection reveals a small load change in the time scale of 10 units or so (i.e., a duration of about $3000sec$) before and after the time index 100; a load change of a much larger time scale and magnitude occurs around the time index 150, and a few other load changes of a smaller time scale occur afterwards. A larger $M$ ignores load changes that occur at the smaller time scale and with a smaller magnitude that are otherwise detected by a smaller $M$.

## 2.7 Summary

Network traffic may fluctuate frequently and often unexpectedly for various reasons such as transitions in user behavior and failure of network elements. Thus, on-line monitoring of traffic load is the basic step towards detecting changes of load and engineering network traffic. Timely detection of such changes in traffic is critical for initiating appropriate traffic engineering mechanisms. The performance of a change detection algorithm depends on the *accuracy* of traffic measurement. But, inspecting *every* packet traversing a link to obtain the *exact* amount of traffic load impairs the processing capacity of a router. Therefore, *sampling* techniques that *estimate* traffic accurately with minimal measurement overhead are needed. Static sampling techniques may result in either inaccurate undersampling or unnecessary oversampling. In this chapter, we proposed an adaptive random sampling technique that *bounds the sampling error* to a pre-specified tolerance level while *minimizing* the number of samples.

We have shown that the minimum number of samples needed to maintain the prescribed accuracy is proportional to the squared coefficient of the variation (*SCV*) of packet size distribution. Since we do not have *a priori* knowledge about key traffic parameters – *SCV* of packet size distribution and the number of packets, these parameters are predicted using AR model. The sampling probability is then determined based on these predicted parameters and thus varied adaptively according to traffic dynamics. From the sampled packets, the traffic load is then estimated. We have also derived a theoretical upper bound on the variance of estimation error that affects the robustness of a change detection algorithm. We have experimented with real traffic traces and demonstrated that the proposed adaptive random sampling is very effective in that it achieves the desired accuracy, while also yielding significant reduction in the fraction of sampled data. As part of our ongoing efforts, we are working on extending the proposed sampling technique to address the problem of flow size estimation.

The time series of traffic loads thus estimated are then analyzed using an on-line, non-parametric change detection algorithm to find non-stationarities. This algorithm detects changes in the estimated traffic loads at some (configurable) operational time scale that is larger than the time scale at which the traffic loads are estimated. We have investigated the impact of sampling error on the performance of change detection algorithm and illustrated the *desirability of controlling* estimation error. We believe that our adaptive random sampling technique combined with an on-line change detection algorithm can enable intelligent traffic control and engineering in a scalable manner.

# References

1. Abilene Internet2. http://abilene.internet2.edu.
2. A. K. Bera and C. M. Jarque. An efficient large-sample test for normality of observations and regression residuals. In *Working Papers in Economics and Econometrics, Australian National University*, volume 40, 1981.
3. D. Berry and B. Lindgren. *Statistics theory and Methods*. Duxbury Press, ITP, 2nd edition, 1996.
4. P. Billingsley. *Probability Measures*. Wiley-Interscience Publication, 1995.
5. CAIDA. The Cooperative Association for Internet Data Analysis. http://www.caida.org.
6. Cisco netflow. http://www.cisco.com/en/US/tech/tk648/tk362/tk812/tech_protocol_home. html.

7. K. Claffy, George C. Polyzos, and Hans-Werner Braun. Application of sampling methodologies to network traffic characterization. In *Proceedings of ACM SIGCOMM*, pages 13–17, San Francisco, CA, September 1993.

8. I. Cozzani and S. Giordano. A measurement based qos evaluation through traffic sampling. In *Proceedings of IEEE SICON*, Singapore, 1998.

9. N. Duffield and M. Grossglauser. Trajectory sampling for direct traffic observation. In *Proceedings of ACM SIGCOMM*, 2000.

10. N. Duffield, C. Lund, and M. Thorup. Charging from sampled network usage. In *Proceedings of ACM SIGCOMM Internet Measurement Workshop*, 2001.

11. C. Estan and G. Varghese. New directions in traffic measurement and accounting. In *Proceedings of ACM SIGCOMM*, 2002.

12. J. Gottman. *Time-series analysis*. Cambridge University Press, 1981.

13. Sprint ATL IPMon project. *http://ipmon.sprint.com*.

14. The Internet Traffic Archive. http://ita.ee.lbl.gov.

15. J. Jedwab, P. Phaal, and B. Pinna. Traffic estimation for the largest sources on a network, using packet sampling with limited storage. Technical Report HPL-92-35, HP Labs Technical Report, 1992.

16. V. Moskvina and A. Zhigljavsky. Change-point detection algorithm based on the singular-spectrum analysis, detection. In *Cardiff University, UK*.

17. C. R. Rao. *Sampling Techniques*. 2nd ed., N.Y., Wiley, 2nd edition, 1973.

18. Cisco sampled netflow. http://www.cisco.com/univercd/cc/tc/doc/product/software/ios120/120newft.

19. A. Sang and S. Q. Li. A predictability analysis of network traffic. In *Proceedings of INFOCOM*, 2000.

20. ACM SIGCOMM. Special Interest Group on Data Communications. http://www.acm.org/sigcomm.

21. W. Stallings. *SNMP, SNMPv2, SNMPv3, and RMON 1 and 2*. Addison Wesley, 3rd edition, 1999.

22. W. Willinger, Murad Taqqu, and Ashok Erramilli. *A Bibliographical Guide to Self-Similar Traffic and Performance Modeling for Modern High-Speed Networks Stochastic Networks: Theory and Applications, Royal Statistical Society Lecture Notes Series*, volume 4. Oxford University Press, 1996.

# Chapter 3
# Flow Characterization and Measurement

**Abstract** Traffic measurement and monitoring are an important component of network management and traffic engineering. With high-speed Internet backbone links, efficient and effective packet sampling techniques for traffic measurement and monitoring are not only desirable, but also increasingly becoming a necessity. Since the utility of sampling depends on the *accuracy* and *economy* of measurement, it is important to *control* sampling error. In this paper we propose and analyze an *adaptive, stratified* packet sampling technique for *flow-level* traffic measurement. In particular, we address the *theoretical and practical issues* involved. Through theoretical studies and experiments, we demonstrate that the proposed sampling technique provides unbiased estimation of flow size with *controllable error bound*, in terms of both packet and byte counts for *elephant* flows, while avoiding excessive oversampling.

**Key words:** adaptive random sampling, flow measurement, packet trace, estimation accuracy, packet count, byte count

## 3.1 Introduction

Traffic measurement and monitoring serve as the basis for a wide range of IP network operations, management, and engineering tasks. Particularly, *flow-level* measurement is required for applications such as traffic profiling, usage-based accounting, traffic engineering, traffic matrix, and QoS monitoring. Traditionally, every packet traversing a measurement point is captured by a router (Figure 3.2) while forwarding it, or by a middlebox [6] (e.g., a measurement probe) attached to a switch interface or a link. With today's high-speed (e.g., Gbps or Tbps) links, such an approach may no longer be feasible. Because flow statistics are typically maintained by *software*, the processing speed cannot match the line speed. Furthermore, the *large number of flows* observed on today's high-speed links introduces scalability issues in traffic measurement. Capturing every packet requires too much *processing*

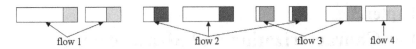

**Fig. 3.1** Packets of flows.

*capacity, cache memory, and I/O and network bandwidth, for updating, storing, and exporting flow statistics records.* Packet sampling has been suggested as a scalable alternative to address this problem. Both the Internet IETF (Internet Engineering Task Force) working groups, IPFIX (IP Flow Information Export) [14] and PSAMP (Packet Sampling) [16], have recommended the use of packet sampling. Static sampling method such as "1 out of $k$" is being used by Cisco and Juniper for high-speed core routers ([15, 18]).

The foremost and fundamental question regarding sampling is its *accuracy*. As stated previously, an inaccurate packet sampling not only defeats the purpose of traffic measurement and monitoring, but worse, can lead to wrong decisions by network operators. Particularly, when it comes to accounting, users would not make a monetary commitment based on erroneous and unreliable data. An important related concern is the efficiency of packet sampling. Excessive oversampling should also be avoided for the measurement solution to be scalable, especially in the presence of high day/night traffic fluctuations, which are well known (see Figure 3.10 for example). Therefore, it is important to *control the accuracy* of estimation in order to *balance the trade-off between the utility and overhead of measurement.* Given the dynamic nature of network traffic, *static* sampling, where a fixed sampling rate is used, does not always ensure the accuracy of estimation, and tends to oversample at peak periods when economy and timeliness are most critical.

Packet sampling for *flow-level measurement* 3.1 is a particularly challenging problem. One issue is the diversity of flows: flows can vary drastically in their volumes. The dynamics of flows is another issue: flows arrive at random times and stay active for a random duration; the rate of a flow (i.e., the number of packets generated by a flow per unit of time) may also vary over time, further complicating the matter of packet sampling.

How can we ensure accuracy of measurement of *dynamic* flows? How many packets does one need to sample in order to produce *flow measurement* with a *pre-specified error bound*? How do we decide on a sampling rate to avoid excessive oversampling while ensuring accuracy? How do we perform sampling procedure and estimate flow volume? How easily can it be implemented at line speed? To answer these questions, we advance a theoretical framework and develop an *adaptive* packet sampling technique using *stratified random sampling.*

The technique is targeted for *accurate* (i.e., with *bounded* sampling errors) estimation of *large* or *elephant* flows based on sampling. That we focus only on large flows is justified by many recent studies ([2, 11, 12]) that demonstrate the prevalence of the "elephant and mice phenomenon" for flows defined at various levels of granularity: a small percentage of flows typically accounts for a large percentage of the total traffic. Therefore, for many monitoring and measurement applications,

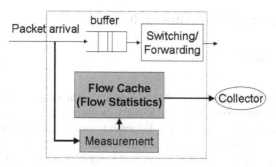

**Fig. 3.2** Flow measurement.

accurate estimation of flow statistics for elephant flows is often sufficient. We employ stratified random sampling to circumvent the issues caused by flow dynamics. Through theoretical analysis, we establish the properties of the proposed adaptive stratified random sampling technique for flow-level measurement. Using real network traffic traces, we demonstrate that the proposed technique indeed produces the desired accuracy of flow volume estimation, while at the same time achieving significant reduction in the amount of packet samples and flow cache size.

The remainder of the chapter is organized as follows. In Section 3.2, we provide an overview of the challenges in packet sampling for flow-level measurement and our proposed approaches. In Section 3.3, we formally state the flow volume estimation problem. We then analyze how sampling errors can be bounded within pre-specified accuracy parameters under dynamic traffic conditions. In Section 3.4, we discuss practical implementation issues involved. Experimental results using network traffic traces are presented in Section 3.5. The chapter is summarized in Section 3.6.

**Table 3.1** Summary of traces used.

| Name | Trace | Date | Avg Load | Duration |
|------|-------|------|----------|----------|
| $\Pi_1$ | OC3 Auck-II | Oct. 2001 | 152Kbps | 4hr |
| $\Pi_2$ | OC3 Tier-1 Backbone | Aug. 2002 | 49.1Mbps | 30min |
| $\Pi_3$ | OC12 Tier-1 Backbone | Aug. 2002 | 43.4Mbps | 30min |
| $\Pi_4$ | OC48 Tier-1 Backbone | Aug. 2002 | 510.9Mbps | 30min |
| $\Pi_5$ | OC12 Tier-1 Backbone | Aug. 2002 | 5.2Mbps | 24hr |
| $\Pi_6$ | OC12 AIX | Oct. 2001 | 21.6Mbps | 90sec |

## 3.2 Challenges and Our Approach

### 3.2.1 Background

A flow is a sequence of packets that share certain common properties (called *flow specification*) and have some temporal locality as observed at a given measurement point (See Figure 3.1 for illustration.). Depending on the application and measurement objectives, flows may be defined in various manners such as source/destination IP addresses, port numbers, protocol, or combinations thereof. They can be further grouped and aggregated into various granularity levels such as network prefixes or autonomous systems. Our analysis, providing bounded accuracy in flow volume estimation, applies to *any* kind of flow definition. For illustrational consistency, in this chapter we present flow statistics and experimental results using flows of *5-tuple* (source/destination IP addresses, port numbers and protocol number) with a *60sec* timeout value as our basic flow definition. The 5-tuple definition is at the finest granularity using packet header traces. The traces used in this study are obtained from both public and commercial OC3, OC12, and OC48 links. The public traces are from CAIDA [5] and the commercial link traces are from the tier-1 ISP backbone network. The trace statistics are listed in Table 3.1.

As illustrated in Figure 3.2, flow measurement in routers works as follows. When a packet arrives, it is classified into a flow. If the flow state is already present in the flow cache, the corresponding flow information, such as flow packet count and byte count, is updated. Otherwise, a new entry is created in the cache. When no new packet arrives within a given timeout period since the arrival of the last packet, this flow is terminated and the flow statistics are exported to a collector entity [14].

Thus, the overhead involved in flow measurement is the following: *Per each packet*, it has to be *classified* into a flow, and its flow information should be either *updated or created*. *Per each flow*, its statistics should be *stored* during its lifetime, and are exported to a collector when it expires or periodically, *consuming bandwidth*. Considering huge number of packets and flows in the Internet backbone traffic, the scalability issue in flow measurement arises from both the number of packets and flows.

### 3.2.2 Observations on Flow Characteristics and the Impact on Packet Sampling

#### 3.2.2.1 Flow Size Diversity

Clearly, flows are quite diverse in their sizes. Note that extremely small flows (e.g., with 10 or fewer packets) may not be detected at all using packet sampling; thus, it would be infeasible to achieve any reasonable degree of accuracy.

Fortunately, for many traffic monitoring and measurement applications, it is sufficient to provide an accurate estimate of flow sizes for only *large* flows. This is due to the fact that the small percentage of large flows typically accounts for a large percentage of total traffic. This is evident in Figure 3.3 where we order the flows based on their packet counts, and plot the cumulative probability, where they account for total traffic (in terms of packet count). We see that less than $10 \sim 20\%$ of the top-ranked flows are responsible for more than 80% of the total traffic different links. The aforementioned phenomenon has been referred to as the "elephants and mice phenomenon" in the literature, and has been observed at various granularities such as point-to-multipoint router level [12], network prefix level [2] and inter-AS level [11]. The observation suggests that meaningful traffic monitoring and measurement objectives (e.g., for traffic engineering or profiling) can often be achieved by concentrating only on a relatively small percentage of large (i.e., elephant) flows.

This motivates us to develop a packet sampling technique to *accurately* estimate *elephant* flows. Such a packet sampling technique reduces the per-packet processing overhead such as classification and flow statistics update.

Figure 3.4 shows the cumulative probability distribution of flow sizes in terms of packet count (i.e., number of packets) for flows in the traces. The majority (80%) of the flows are small (e.g., with 10 or fewer packets), while a small percentage of them are large (e.g., with more than $10^5$ or $10^6$ packets). This means that many small flows may not be detected by packet sampling, leading to a reduction in flow cache size. Thus, a sampling technique would relieve the overhead related to both per-packet and per-flow.

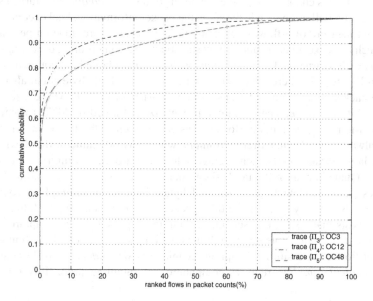

**Fig. 3.3** Elephants-mice behavior.

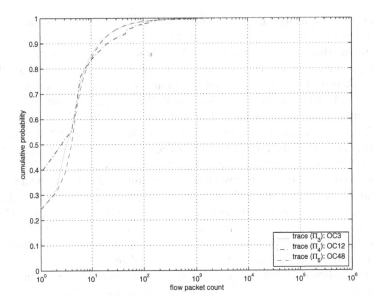

**Fig. 3.4** Many small flows.

An elephant flow or a large flow can be defined in various ways, e.g., in terms of a packet count, a byte count (i.e., number of bytes), or even some measure of burstiness. In this chapter, we define it in terms of a *packet count*.[1] In return, a sampling decision will not be made based on the packet information. Packet count is an important measure of a flow, since many tasks in a router are done on per-packet basis such as packet classification, flow statistic update, and routing decision. A set of flows with a large packet count contains flows with a large byte count, as is evident in Figure 3.5, where we observe that flows with a large byte count also have a large packet count. This is because a flow consists of packets whose sizes cannot be arbitrarily large. The maximum packet size is limited by MTU (Maximum Transmission Unit) on a link.[2] However, as can be seen in the figure, the opposite is not always true. As an extreme example, while network game traffic flows tend to have large packet counts, their average byte counts are small. Empirical measurement of 'Quake' client traffic in [4] shows that the mean packet size is around only 24 bytes (standard deviation of around 1 byte). Thus, the resulting byte count of such flows is not very large, while their packet count ranges from 14000 to 39000. This suggests that a sampling scheme without using packet size can capture flows with large byte count. Note that by relying only on packet count in the definition of elephant flows, the resulting sampling technique is *size-independent*. In contrast, *size-dependent sampling* requires computation of sampling probability for each

---

[1] We will give a formal definition of an elephant flow used in this work in the next section.

[2] Note that the linear lines in Figure 3.5 are due to the dominant packet sizes (40, 570, and 1500 bytes).

object (either a packet or a flow) [8, 10], thus increasing the per-object processing overhead.

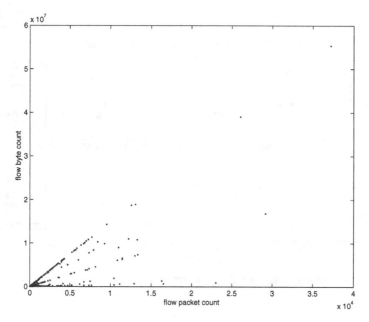

**Fig. 3.5** Correlation of flow byte count and packet count (trace $\Pi_1$).

### 3.2.2.2 Flow Dynamics

So far, we have observed flow statistics, which beneficially influence on packet sampling. Now we discuss issues that challenges the problem of flow measurement.

Flows are dynamic, in their arrival time, duration, and rate over time (the number of packets/bytes generated by a flow per unit of time); flows arrive at random times, and stay active for random durations. Furthermore, the rate of a flow varies during the flow duration. In Figure 3.6, we illustrate several top-ranked flows from a trace ($\Pi_1$). The flows are ranked in packet count over the trace. The top plot shows the average packet size of each flow over a 5 minute interval, and the middle and the bottom plot depict the byte count and the packet count of each flow over 5 minute intervals, respectively. Since the average packet sizes are different *among* flows, flow ranks in packet count may not be equivalent to ranks in byte count. On the other hand, packet size *within* a flow remains pretty much the same over its life time.

As pointed out in the previous chapter and 2, in order to achieve both sampling accuracy and efficiency at the same time, it is important to *adapt* the sampling

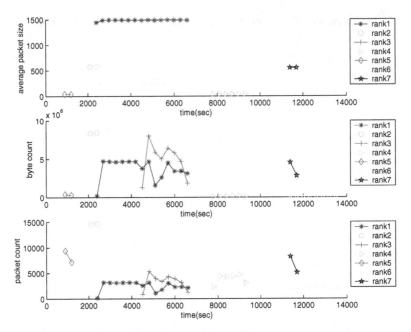

**Fig. 3.6** Flow dynamics (ranked in packet count) (trace $\Pi_1$).

rate according to changes in the traffic. Under dynamically changing traffic conditions, *static* sampling rate may lead to *inaccurate undersampling* or *excessive oversampling*. Such problems become more acute considering long-term daily scale, where the daytime traffic rate differs significantly from the nighttime rate as shown Figure 3.10. However, it is hard to decide a *sampling interval* and an optimal *sampling probability* on an interval, due to the dynamic flow arrival and duration.

In addition, the byte/packet counts over time vary a lot within a flow. When a flow rate is used in defining an elephant flow, a flow's classification (elephant or mouse) may change over time, as its rate changes. Thus, the changing rate within a flow makes it difficult to define elephant flows.

We outline how we tackle the above issues in the next subsection.

### 3.2.3  Our Approach

For accuracy and efficiency of sampling, a sampling interval should be determined, for which a sampling rate is adjusted in accordance with the changing traffic condition. Because flows are dynamic in their arrival time and active duration (as seen in Figure 3.6), it is very hard to define a sampling frame (i.e., a sampling interval) that is valid for *all* elephant flows. We tackle this problem by using *stratified*

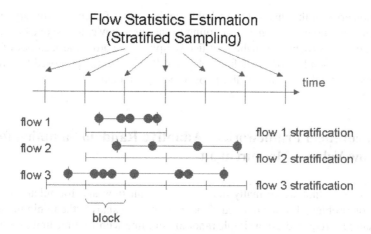

**Fig. 3.7** Stratified random sampling.

sampling. As illustrated in Figure 3.7, it uses *predetermined, non-overlapping* time blocks called *strata*. For each block, it samples packets with the same probability (i.e., via simple random sampling). At the end of each block, flow statistics are estimated. Then, naturally, a flow's volume is summarized into a single estimation record at the end of the last time block enclosing the flow. Notice that from each flow's point of view, its duration is divided or *stratified* in a fixed time. The predetermined time blocks enable us to estimate flow volume, without knowing dynamic flow arrival times and their durations. The sampling rate is adjusted according to dynamical traffic changes.

We classify flows based on *a proportion of packet count over a time interval that encompasses the flow duration*; i.e., a flow is referred to as an *elephant* flow in our study if its packet count proportion is larger than a pre-specified threshold (for example, 1% of the total traffic).

The proposed definition of elephant flow captures large packet and byte count flows as well as high rate (or bursty) flows. Furthermore, it removes the difficulty dealing with intra-flow rate variability. Flows with *packet count over a certain threshold* capture natually high byte count flows as well as high packet count flows, as discussed before. The *proportion* tells us how *bursty* a flow is or the rate of an (entire) flow, compared to other simultaneous flows. Since the proportion is defined over *the time interval enclosing a flow*, it eliminates the intra-flow rate fluctuation issue. We will provide a more rigorous definition of an elephant flow in Section 3.3.

A time *block* is the minimum time scale over which an elephant flow (packet count proportion) is identified. It is also the minimum time scale over which the sampling rate can be adjusted. As will be shown in Section 3.3, in order to achieve a desired accuracy, at least a certain number of packets must be sampled during the sampling frame that encompasses an elephant flow duration. The sampling rate is set to collect the required number of samples in a block in order to bound the estimation error of the smallest (threshold) elephant flow. Given the arbitrary length of the elephant flow duration, the sampling frame for a flow could be one block or a series of consecutive blocks in the stratified sampling. We prove the accuracy of flow estimation is bounded for the defined elephant flows with the proposed technique, regardless of the flow's rate variability over multiple blocks.

## 3.3 Theoretical Framework of Adaptive Random Sampling for Flow Volume Measurement

In this section, first we formally define an elephant flow and formulate the flow estimation problem. For the defined elephant flows, we analyze the minimum number of samples required using simple random sampling within a time unit in order to bound sampling errors. We then describe how to determine sampling probability and how the accuracy is achieved for flows of arbitrary lengths using stratified random sampling. Finally, we establish the statistical properties of the proposed technique.

### 3.3.1 Elephant Flow Definition and Problem Formulation

In this chapter, a flow is referred to as an elephant flow if its packet count proportion is larger than a pre-specified threshold over a flow encompassing time interval (for example, 0.1%). For those elephant flows, the proposed sampling technique estimates *flow packet count as well as byte count with controlled accuracy*.

First, we formally give a definition of an elephant flow used in this chapter.

**Definition 1 (Elephant Flow)** *Consider a discretized time interval that contains an entire duration of flow $f$. Suppose the interval consists of $L$ consecutive (time) blocks where $m_i$ packets are seen in block $i$ ($i = 1 \ldots L$). Let $m^f$ packets belong to flow $f$ out of total $m$ packets. If the proportion of flow packet count $p^f$ is greater than a threshold $p^\theta$, then we call the flow an* elephant.

$$\frac{m^f}{m} = \frac{\sum_{h=1}^{L} m_h^f}{\sum_{h=1}^{L} m_h} = p^f \geq p^\theta \tag{3.1}$$

Determining flow rate over a certain time scale of interest is reasonable for practical issues. A straightforward computation of flow rate, i.e., flow size divided by its duration, may not be meaningful, particularly for very short flows. For example, the rate for single packet flows is not well-defined since its duration is considered to be zero. On the other hand, flows with two packets sent back-to-back would give the highest flow rate which is equivalent to a link rate. For long lived flows, classifying flows over a duration that might be slightly larger than its actual duration has only minimal impact on the class characteristic.

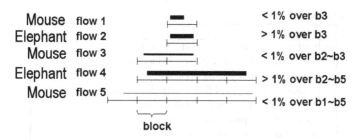

**Fig. 3.8** Elephant flow used in this chapter.

The time scale over which flows are classified can be determined by a certain engineering purpose. Flow classification is relative to the total traffic during the flow's life time. The online identification of a flow as an elephant or a mouse can be done by just keeping one counter of the total packet for blocks of a flow duration. When the flow expires, the packet count proportion of the flow over the total packet counts during the blocks indicates whether the flow is an elephant or not. If it is indicated as an elephant, the flow volume estimation should be accurate with the pre-specified error bound.

Our objective is to bound the relative error of packet count estimation, $\hat{m}^f$ and byte count estimation, $\hat{v}^f$ for the elephant flows, i.e., given the *prescribed* error tolerance level, $\{\eta, \varepsilon\}$, (where $(1 - \eta)$ and $\varepsilon$ are referred as *reliability* and *precision* respectively, and $0 \leq \eta \leq 1$), flow packet count and byte count estimation error have to be bounded respectively as

$$Pr\left\{\left|\frac{\hat{m}^f - m^f}{m^f}\right| > \varepsilon\right\} \leq \eta, \, Pr\left\{\left|\frac{\hat{v}^f - v^f}{v^f}\right| > \varepsilon\right\} \leq \eta \qquad (3.2)$$

where $p^f \geq p^\theta$ for flow $f$. In other words, we want the relative error in flow volume estimation using random sampling to be bounded by $\varepsilon$ with a high probability $1 - \eta$. Given this formulation of the bounded error sampling problem, the question is *what is the minimum number of packets that must be sampled randomly so as to guarantee the prescribed accuracy for diverse and dynamic flows.* We address this question in the following subsection.

We have chosen to bound relative error, since it gives generic accuracy regardless of load, link or characteristic. However, we will also discuss about bounding absolute error at the end of this section. The notations used here are summarized in Table 3.2.

**Table 3.2** Notation.

|     | Explanation |
| --- | --- |
| $m_h$ | total number of arriving packets in block $h$ |
| $n_h$ | total number of sampled packets in block $h$ |
| $p^f$ | proportion of flow $f$ in packet counts |
| $\hat{p}^f$ | estimated proportion of flow $f$ in packet counts (r.v.) |
| $n^f$ | number of sampled packets of flow $f$ (r.v.) |
| $m^f$ | total number of packets of flow $f$ |
| $\hat{m}^f$ | estimated packet count of flow $f$ (r.v.) |
| $v^f$ | byte count of flow $f$ |
| $\hat{v}^f$ | estimated byte count of flow $f$ |
| $S^f$ | squared coefficient of variation (SCV) of packet sizes of a flow $f$ |
| $p^\theta$ | elephant flow threshold (in packet count proportion) |
| $S^\theta$ | threshold in SCV of elephant flow packet sizes |

## 3.3.2 Required Number of Samples

Our approach and analysis framework are based on random sampling. The assumptions we make in the analysis are that the sample size $n$ is reasonably large ($> 30$ packets) and the population size $m$ is large enough compared to the sample size ($m \gg n$) so that the sampling fraction is small. Then, the sampling distribution of the sample mean for *random samples* has a normal distribution with the mean $\mu$ and standard deviation $\frac{\sigma}{\sqrt{n}}$, *regardless of the distribution of population*, from the Central Limit Theorem. $\mu$ and $\sigma$ are the population mean and the standard deviation, respectively. Recall that the requirement of samples being i.i.d (independent and identically distributed) for the condition of the theorem is simply achieved by *random* sampling from the *common* population.[3]

We first derive the required number of samples to provide the pre-specified accuracy using simple random sampling. We then explain how the accuracy is achieved for flows active over multiple blocks in the next subsection.

### 3.3.2.1  Flow Packet Count Estimation

Using a simple random sampling, a flow packet count is estimated as follows. Consider a unit time interval that contains an *entire duration* of flow $f$, in which $m$ packets are seen. From these, $n$ packets are *randomly sampled* ($n < m$), and $n^f$ packets belong to flow $f$. Then the packet count of flow $f$, $m^f$ is estimated by $\hat{m}^f$ using the sample proportion $\hat{p}^f$:

---

[3] It is important to understand that *randomizing eliminates correlation*. For example, in [9], the randomizing technique is used to destroy correlation for the purpose of investigating the impact of long range dependence on the queueing performance.

$$\hat{m}^f = m \cdot \frac{n^f}{n} = m \cdot \hat{p}^f \qquad (3.3)$$

A proportion may be considered to be a special case of the mean where a variable $Y$ takes on only the values 0 and 1. For example, suppose we wish to find the proportion of a particular flow $f$. Let there be $m$ packets, and let $Y_i = 1$ if the $i$th packet belongs to the flow $f$, and $Y_i = 0$ otherwise. Then the number of packets belonging to the flow $f$ is

$$m^f = \sum_{i=1}^{m} Y_i \qquad (3.4)$$

The flow proportion of packets is computed by the total packet count during the interval

$$p^f = \frac{m^f}{m} = \frac{\sum_{i=1}^{m} Y_i}{m} \qquad (3.5)$$

Let $Y_1, Y_2, \ldots, Y_n$ be $n$ random samples, and the $n^f$ packets of them belong to flow $f$. The sample proportion of flow $f$ is therefore defined as

$$\hat{p}^f = \frac{n^f}{n} = \frac{\sum_{j=1}^{n} Y_j}{n} \qquad (3.6)$$

Within a time block, a simple random sampling in which a sampling probability is fixed is used. Then, from the Central Limit Theorem of *random samples* [1], as the sample size $n \to \infty$, the sample mean $\hat{p}^f$ approaches the population mean $p^f$ and variance $\sigma_{\hat{p}^f}^2 = p^f(1 - p^f)/n$ regardless of the distribution of population. Thus, the sample proportion can be written with its mean and variance,

$$\hat{p}^f \approx p^f + \frac{\sqrt{p^f(1 - p^f)}}{\sqrt{n}} Y_p \qquad (3.7)$$

where $Y_p$ is a random number of a standard normal distribution ($\sim N(0, 1)$) and the subscript $p$ stands for packet count.

Now Eq. (3.2) can be rewritten as follows:

$$Pr\left\{ \left| \frac{m\hat{p}^f - mp^f}{mp^f} \right| > \varepsilon \right\} = Pr\left\{ \left| \frac{\hat{p}^f - p^f}{\sigma_{\hat{p}^f}} \right| > \frac{p^f\sqrt{n}\varepsilon}{\sqrt{p^f(1 - p^f)}} \right\} \qquad (3.8)$$

$$\approx 2\left( 1 - \Phi\left( \frac{\sqrt{p^f}\sqrt{n}\varepsilon}{\sqrt{(1 - p^f)}} \right) \right) \le \eta$$

where $\Phi(\cdot)$ is the cumulative distribution function (c.d.f) of the standard normal distribution.

By solving the inequality in Eq. (3.9) with respect to $n$, we can derive the minimum required number of samples $n^{*,p}$ to estimate flow packet count within the given error tolerance level

$$n \geq n^{*,p} = \left\lceil z_p \cdot \left( \frac{1 - p^f}{p^f} \right) \right\rceil \tag{3.9}$$

where $z_p = \left( \frac{\Phi^{-1}(1 - \eta/2)}{\varepsilon} \right)^2$.

Notice that with the elephant threshold of packet count proportion $p^\theta$, $\left( \frac{1 - p^\theta}{p^\theta} \right)$ can be set as a constant $C_\theta = \left( \frac{1 - p^\theta}{p^\theta} \right)$. Then,

$$n \geq n^{*,p} = \lceil z_p \cdot C_\theta \rceil \tag{3.10}$$

With at least $n^{*,p}$ number of random samples, simple random sampling can provide *pre-specified accuracy* $\{\eta, \varepsilon\}$ for *any* flows whose proportion is larger than a pre-defined elephant threshold $p^\theta$.

Eq. (3.10) concisely *relates* the *minimum number of packet samples* to the estimation *accuracy* and the *elephant flow threshold*. Moreover, given accuracy and the elephant flow threshold, it shows that the amount of measurement needed remains *constant* regardless of the traffic fluctuation.

### 3.3.2.2 Flow Byte Count Estimation

For the defined elephant flows, we also aim to measure flow byte count accurately, in addition to flow packet counts. The actual byte count of a flow $f$ is expressed as follows:

$$v^f = m^f \mu^f = mp^f \mu^f \tag{3.11}$$

where $\mu^f$ is the actual average packet size of flow $f$. Similarly the estimated flow byte count $\hat{v}^f$ is

$$\hat{v}^f = \hat{m}^f \hat{\mu}^f = m\hat{p}^f \hat{\mu}^f \tag{3.12}$$

where $\hat{\mu}^f$ is the estimated average packet size of flow $f$.

Notice that two levels of uncertainties are involved for flow byte count estimation, namely the estimations of flow proportion and flow's average packet size.

The flow byte count estimation can be quantified with the help of the following two lemmas, which are the consistency of sample proportion, and an extension of the Central Limit Theorem for a sum of a random number of random variables, respectively.

**Lemma 1** $\frac{n^f}{n \cdot p^f} \to 1$ *almost surely as* $n \to \infty$ *by the strong law of large numbers.*

**Lemma 2 (p369, problem 27.14 in [3])** *Let* $X_1, X_2, \ldots$ *be independent, identically distributed random variables with mean* $\mu$ *and variance* $\sigma^2$, *and for each positive n, let* $F_n$ *be a random variable assuming positive integers as values; it does not need to be independent of the* $X_m$'s. *Let* $W_n = \sum_{i=1}^{F_n} X_i$. *Suppose then as* $n \to \infty$, $\frac{F_n}{n}$ *converges to 1 almost surely. Then as* $n \to \infty$,

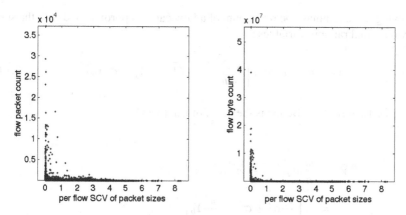

**Fig. 3.9** Flow packet count and byte count vs. SCV.

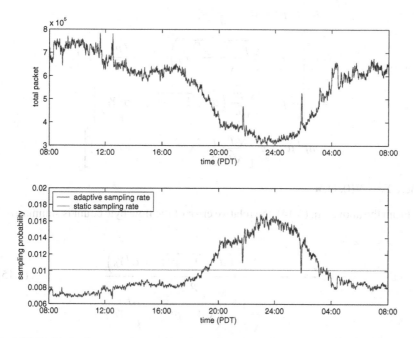

**Fig. 3.10** Traffic load, total packet count and sampling probability (trace $\Pi_5$, $\{\eta, \varepsilon\} = \{0.1, 0.1\}$, $\{p^\theta, S^\theta\} = \{0.01, 0.2\}$).

$$\frac{W_n - F_n \mu}{\sigma \sqrt{n}} \tag{3.13}$$

*converges in distribution to a $N(0, 1)$ random variable.*

Applying these lemmas, the byte count of a flow can be approximated with the sum of two normal random variables as

$$\hat{v}^f = mp^f\mu^f + m\left[\frac{\sqrt{p^f}}{\sqrt{n}}\left(\mu^f\sqrt{1-p^f}Y_p + \sigma^f Y_b\right)\right] \tag{3.14}$$

where $Y_b, Y_p \sim N(0,1)$. Below is the proof of Eq. (3.14).

*Proof.*

$$\hat{V}^f = \frac{m}{n}n^f\hat{\mu}^f \approx \frac{m}{n}\left[n^f\mu^f + \sigma^f\sqrt{np^f}Y_b\right]$$

$$= m\left[\frac{n^f}{n}\mu^f + \sigma^f\frac{\sqrt{np^f}}{n}Y_b\right]$$

$$= m\left[\hat{p}^f\mu^f + \sigma^f\frac{\sqrt{p^f}}{\sqrt{n}}Y_b\right]$$

$$= m\left[\left(p^f + \frac{\sqrt{p^f(1-p^f)}}{\sqrt{n}}Y_p\right)\mu^f + \frac{\sqrt{p^f}}{\sqrt{n}}\sigma^f Y_b\right]$$

$$= m\left[p^f\mu^f + \frac{\sqrt{p^f}}{\sqrt{n}}\left(\mu^f\sqrt{1-p^f}Y_p + \sigma^f Y_b\right)\right]$$

$$= mp^f\mu^f + m\left[\frac{\sqrt{p^f}}{\sqrt{n}}\left(\mu^f\sqrt{1-p^f}Y_p + \sigma^f Y_b\right)\right]$$

where $Y_b \sim N(0,1)$. ■

From the above Eq. (3.14), the relative error of the flow byte count is summarized as

$$\frac{\hat{v}^f - v^f}{v^f} = \frac{\frac{\sqrt{p^f}}{\sqrt{n}}\left(\mu^f\sqrt{1-p^f}Y_p + \sigma^f Y_b\right)}{p^f\mu^f} \tag{3.15}$$

$$= \frac{1}{\sqrt{np^f}}\left(\sqrt{1-p^f}Y_p + \frac{\sigma^f}{\mu^f}Y_b\right)$$

$$\sim N\left(0, \frac{1-p^f + S^f}{np^f}\right)$$

Then, the required number of samples for the flow byte count estimation can be obtained similarly to the flow packet count estimation,

$$n \geq n^{*,b,f} = \left\lceil z_p \cdot \left(\frac{1-p^f + S^f}{p^f}\right)\right\rceil \tag{3.16}$$

where $S^f = (\sigma^f/\mu^f)^2$ is the *squared coefficient of variation* (SCV) of packet sizes of flow $f$.

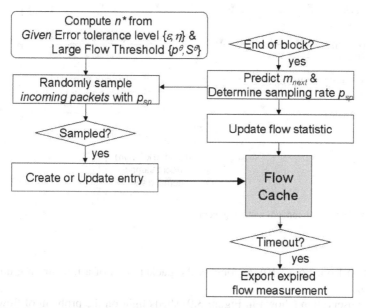

**Fig. 3.11** Adaptive random sampling for flow volume measurement.

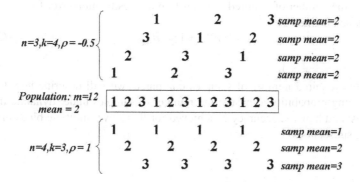

**Fig. 3.12** Systematic sampling on periodic population.

Eq. (3.16) reveals that the required number of samples for a flow byte count estimation is related to the *variability of packet sizes of a flow* as well as the packet count proportion and accuracy. It also tells that a larger number of samples are needed for flow byte count estimation compared with the one for flow packet count

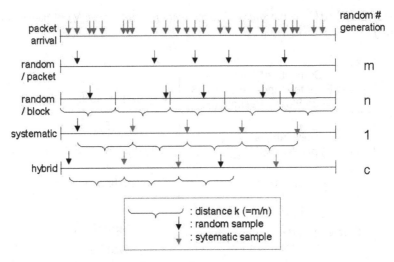

**Fig. 3.13** Trade-off in random number generation.

estimation from Eq. (3.10), as long as the packet sizes of a flow are not uniform $(S^f > 0)$.

Our observation (shown in Figure 3.9) sheds light on the problem of flow byte count estimation. Even though the variability of packet sizes (SCV) of a flow ranges widely in general (from 0.00007 to 8(!)), it is very limited for large flows. This means large flows tend to have packets of similar sizes. One can effectively give a reasonable bound on the SCV of elephant flows, around 0.2($< 1$) for example. Therefore, the number of required samples to bound estimation error for flow byte count can be obtained by

$$n \geq n^{*,b} = \lceil z_p \cdot B_\theta \rceil \tag{3.17}$$

where $B_\theta = \left( \frac{1 - p^\theta + S^\theta}{p^\theta} \right)$.

With the required number of samples computed, we will describe how to decide sampling probability to ensure the number of samples under dynamic traffic conditions, and how the accuracy is achieved for flows over multiple blocks in the following subsections.

### 3.3.3 Optimal Sampling Probability and Prediction of Total Packet Count

The optimal sampling probability of a block to produce $n^*$ samples would be

$$p_{sp} = \frac{n^*}{m_h} \tag{3.18}$$

where $m_h$ is the total number of packets in a block $h$. $n^*$ can be $n^{*,p}$ only for flow packet count or $n^{*,b}$ for flow byte count as well.

In any case, we cannot accurately choose the sampling rate when the population size (total packet count of the observation time block) is unknown. We can compute the sampling probability at the beginning of a block by predicting the total packet count. We employ an the AR (Auto-Regressive) model for predicting the total traffic packet count $m$, as compared to other time series models, since it is easier to understand and computationally more efficient. In particular, using the AR model, the model parameters can be obtained by solving a set of simple linear equations [13], making it suitable for *online implementation*.

We will now briefly describe how the total packet count $m_h$ of the $h$th block can be estimated, based on the past packet counts using the AR($u$) model. Using the AR($u$) model [13], $m_h$ can be expressed as

$$m_h = \sum_{i=1}^{u} a_i m_{h-i} + e_h \qquad (3.19)$$

where $a_i$, $i = 1, \ldots, u$, are the model parameters, and $e_h$ is the *uncorrelated* error (that we refer to as the *prediction error*).

The model parameters $a_i$, $i = 1, \ldots, u$, can be determined by solving a set of linear equations in terms of $v$ past values of $m_i$'s, where $v \geq 1$ is a configurable parameter independent of $u$, and is typically referred to as the memory size.

Let $\hat{m}_h$ denote the *predicted* packet count of the $h$th block. Using the the AR($u$) prediction model, we have

$$\hat{m}_h = \sum_{i=1}^{u} a_i m_{h-i}. \qquad (3.20)$$

In predicting the *total* packet count $m$, we assume the *actual* packet count of a block is known at the end of the block. Note that having the actual total packet count is reasonable to assume in current routers and does not change the nature of the adaptive random sampling technique we propose.

Using the AR prediction model, at the end of each block, the model parameters ($a_i$) are computed [13]. The complexity of the AR prediction model parameter computation is only $O(v)$ where $v$ is the memory size. Through empirical studies, we have found that AR(1) with a small memory size (around 5) is sufficient to yield a good prediction.

For the currently active flows, their statistics are updated using the sampling rate at the end of a block $h$ as follows:

$$\hat{m}_h^f = \hat{m}_{h-1}^f + \frac{m_h}{n_h} \hat{n}_h^f \qquad (3.21)$$

$$\hat{v}_h^f = \hat{v}_{h-1}^f + \frac{m_h}{n_h} \hat{n}_h^f \hat{\mu}_h^f \qquad (3.22)$$

Figure 3.11 shows the flow chart of the adaptive random sampling procedure.

### 3.3.4 Accuracy of Stratified Random Sampling: Statistical Properties

In our proposed sampling method, we collect an *equal* number of random samples ($n^*$) for *each* stratum on average. Consider a flow whose enclosing interval consists of an $L$ number of blocks. Then, *from the flow's point of view, $n^* \cdot L$ packets are sampled for the $L$ blocks and for each block a simple random sampling is used that is equivalent to a *stratified random sampling with an equal number of samples per stratum*. In the previous subsection, we have shown that simple random sampling with $n^*$ samples provides the prescribed accuracy for the estimation of flows whose duration falls within a block. Here, we first explore statistical properties of the stratified random sampling. Then we show how a stratified random sampling with an equal number of $n^*$ samples per stratum also gives at least the prescribed accuracy for flows which live for $L$ blocks.

Stratified random sampling is known to provide *unbiased estimators* for the population mean, total, and proportion, in that their expectations are equal to the values of population ($E(\hat{p}^f) = p^f, E(\hat{v}^f) = v^f$). The technique is also *consistent*, since the estimation of stratified random sampling approaches the population parameter as the number of samples increases. i.e., $\hat{p}^f \to p^f$ as $n \to \infty$ (or $m$).

*Efficiency* of a sampling describes how closely a sampling distribution is concentrated around the value of the population (population parameter). For consistent estimators, efficiency can be measured by the *variance*, where a smaller variance is preferred. An estimator of a smaller variance would give a more *accurate* estimation, given the same number of samples. Mean square error (MSE) is a frequently used metric to compare estimators. Let $X$ be a random variable of the population (in general) and $\hat{X}$ be the estimated mean of the population. (In case of proportion, $X$ takes on 1 if a packet belongs to flow $f$ and 0 otherwise.) As in the following equation, the variance itself becomes MSE for an unbiased estimator.

$$MSE(\hat{X}) = E(\hat{X} - \mu)^2 = Var(\hat{X}) + bias^2 \qquad (3.23)$$

Thus it is important to study variance carefully. Notice that the analysis and the required number of samples in the previous subsections are based on simple random sampling. We first compare the variance of proposed stratified random sampling with simple random sampling. The variance of *total* estimation ($\hat{m}^f$ or $\hat{v}^f$) is easily found by using the results of a variance of *mean* estimation ($\hat{p}^f$ or $\hat{\mu}^f$). For example,

$$Var(\hat{m}^f) = Var(m\hat{p}^f) = m^2 Var(\hat{p}^f) \qquad (3.24)$$

Hence, we simplify the discussion of the variance to a *mean* estimation.

The variance of simple random sampling with $n$ samples is

$$Var(\hat{X}_{sim,n}) = \frac{\sigma^2}{n} \qquad (3.25)$$

where $\sigma^2$ is the population variance [20]. Thus, the accuracy (or variance) of a simple random sampling depends on the variance of the actual population ($\sigma^2$) and a sample size ($n$). The following lemma states that our proposed sampling bounds the variance of relative error by the pre-specified accuracy parameters.

**Lemma 3** *Using $n^*$ random packets, the variance of the relative error in estimating flow packet count and byte count is bounded above by the pre-specified accuracy:*

$$Var\left(\frac{\hat{m}^f - m^f}{m^f}\right) \leq \frac{1}{z_p} \tag{3.26}$$

$$Var\left(\frac{\hat{v}^f - v^f}{v^f}\right) \leq \frac{1}{z_p} \tag{3.27}$$

*where* $z_p = \left(\frac{\Phi^{-1}(1-\eta/2)}{\varepsilon}\right)^2.$

*Proof.*

$$Var\left(\frac{\hat{m}^f - m^f}{m^f}\right) = Var\left(\frac{\hat{p}^f - p^f}{p^f}\right) = \frac{Var\left(\hat{p}^f\right)}{p^{f2}} = \frac{p^f(1-p^f)}{np^{f2}} = \frac{(1-p^f)}{n} \tag{3.28}$$

where $n \geq n^{*,p} = \left\lceil z_p \cdot \left(\frac{1-p^f}{p^f}\right)\right\rceil$.

Thus, $Var\left(\frac{\hat{m}^f - m^f}{m^f}\right) \leq \frac{1}{z_p}.$

Similarly,

$$Var\left(\frac{\hat{V}^f - V^f}{V^f}\right) = \frac{1 - p^f + S^f}{np^f} \tag{3.29}$$

where $n \geq n^{*,b} = \left\lceil z_p \cdot \left(\frac{1-p^f+S^f}{p^f}\right)\right\rceil$

Hence, $Var\left(\frac{\hat{V}^f - V^f}{V^f}\right) \leq \frac{1}{z_p}$ ∎

Now, we consider flows with arbitrary duration that stay active $L(\geq 1)$ blocks. We establish the following theorem to show that the proposed stratified random sampling provides the pre-specified error tolerance.

**Lemma 4** *The variance of stratified random sampling with an equal number of $n^*$ samples for each L strata is smaller than the variance of simple random sampling with n samples.*

$$Var(\hat{X}_{str(eq),n^*L}) \leq Var(\hat{X}_{sim,n^*}) \tag{3.30}$$

*Proof.*

$$Var(\hat{X}_{str,n \cdot L}) = \frac{L}{m^2} \frac{\sum_{i=1}^{L} m_i^2 \sigma_i^2}{nL}$$

$$= \frac{1}{mn} \sum_{i=1}^{L} m_i \sigma_i^2 \cdot \frac{m_i}{m}$$

$$\leq \frac{1}{mn} \sum_{i=1}^{L} m_i \sigma_i^2$$

$$\leq Var(\hat{X}_{sim,n})$$

where $\sigma_i^2$ is a population variance in block $i$. ∎

This means that the accuracy in estimation of a flow with arbitrary duration satisfies the given bound with the proposed stratified random sampling, where $n^*$ samples are collected at each time block.

Another important property to consider for flow is about aggregation. Flow statistics may further be *aggregated into a bigger flow* later for different engineering purposes. It can be easily shown that the *accuracy* of the estimation is *conserved* for aggregated flows.

Stratified random sampling gives smaller variance of estimation (better accuracy) than simple random sampling, when the variances within blocks are small compared to the variance for all the intervals. Given a number of samples $n$, stratified random sampling increases the accuracy, when samples are selected *proportionally* to the population size in the stratum [20]. i.e.,

$$Var(\hat{X}_{str(prop),n}) \leq Var(\hat{X}_{sim,n}) \tag{3.31}$$

If we additionally update flow statistics within a block *without* changing sampling probability, it turns out to be a stratified random sampling. Then the sample sizes in strata (or subblock) become proportional to the population size. Therefore, one may update flow statistics more often than once per block to increase an accuracy of estimation.

### 3.3.5  Bounding Absolute Error

So far we described bounding *relative* error in estimating flow packet and byte count. We have derived the required sample size by linearly separating accuracy parameter from traffic parameter as

$$n \propto \text{accuracy} \cdot \text{traffic parameter} \tag{3.32}$$

For the objective of *absolute* error bound

$$Pr\{|\hat{m}^f - m^f| > \varepsilon\} \leq \eta \text{ or } Pr\{|\hat{v}^f - v^f| > \varepsilon\} \leq \eta \tag{3.33}$$

the required sample size can be obtained by a similar analysis used in bounding relative error. Then, the required number of samples can be derived as

$$n \geq n^{*,p} = z_p \cdot p^f(1-p^f) \cdot m^2 \text{ or} \tag{3.34}$$

$$\tag{3.35}$$

$$n \geq n^{*,b} = z_p \cdot \left[ p^f(1 - p^f)\mu^f + \sigma^f \right] \cdot m^2 \tag{3.36}$$

for flow packet and byte count estimation respectively. It can be summarized in the following form:

$$n \propto \text{accuracy} \cdot \text{traffic parameter} \cdot m^2 \tag{3.37}$$

Assuming $p^f < 0.5$, the required sample size depends on the *maximum* flow proportion whose estimation error should be bounded. Suppose $n^{*,p}$ is computed using $p^\theta$. Since the accuracy is not guaranteed for flows whose proportions are larger than $p^\theta$, the largest flow proportion should be known ahead. Still, sampling probability is much higher than the one for relative error bound in general, since it is quadratically proportional to the total packet count ($m^2$). For flow byte count estimation, two parameters of a flow - average packet size of a flow $\mu^f$ and its variance $\sigma^f$ - should be known as opposed to one parameter of flow SCV for relative error. Furthermore, unlike SCV, the mean packet size of a flow $\mu^f$ varies widely among all the flows as observed in Figure 3.6. Using maximum packet size (1500 bytes for example) would give a *very* large number in the required number of samples (often as many as total packet count), resulting in oversampling for many elephant flows with a smaller average packet size.

Therefore, bounding relative error is more practical, and it is suitable for types of applications such as traffic profiling and engineering, where flows responsible for most of the traffic are of interest.

## 3.4 Practical Considerations

In this section, we discuss the issues involved in the implementation of the proposed sampling technique. We first discuss how to determine the flow timeout value and its impact on the performance of our method. Then we discuss how to reduce the overhead of random number generation.

### 3.4.1 Flow Timeout Value

The choice of timeout values may change the flow statistics for a given traffic even with full measurement (i.e., no sampling). While a large timeout value leads to maintaining an unnecessarily large number of flow states, a small timeout may break otherwise large, long-lived flows into smaller flows. The impact of timeout values on flow statistics and a performance trade-off were studied in [7, 17].

With the introduction of sampling, the timeout value should be adjusted appropriately due to increased inter-packet arrival time of a flow. Suppose $TO_{full}$ is used for a flow definition when no sampling is used. With random sampling, the intra-flow inter-packet arrival time of samples is increased by the inverse of the sampling rate on average ($p_{avg}$). Thus, one can use the timeout value under sampling as

$$TO_{samp} = \frac{TO_{full}}{p_{avg}} \tag{3.38}$$

However, in the proposed adaptive random sampling, the sampling rate changes adaptively over time. So we exponentially average the sampling rate as follows:

$$p_{avg,i} = \alpha p_i + (1 - \alpha) p_{avg,i-1} \tag{3.39}$$

where $p_i$ is the sampling probability of a block $i$ and $p_{avg,i}$ is the averaged sampling probability in block $i$ to be used for a timeout value. Given a timeout value for full measurement ($60sec$), we rarely observe truncation of elephant flows throughout the experiments. It is because from our definition, an elephant flow is expected to have a higher packet rate on average over the flow's duration, thus making it *less sensitive* to a timeout value.

### 3.4.2 Utility of Systematic Sampling and Random Number Generation

Systematic sampling is a popular sampling design employed in Cisco and Juniper routers ([15, 18]). In general, 1-out-of-$k$ systematic sampling involves random selection of one element from the first $k$ elements, and selection of every $k$th element thereafter, requiring only one random number generation and a counter. Random sampling involves a random number generation per packet. Even though modern routers already have the feature implemented for a mechanism such as RED (Random Early Detection), with a choice of a sampling rate from our analysis, one may want to consider using systematic sampling for simplicity. However, understanding accuracy of systematic sampling has to precede its utility.

The performance of systematic sampling can be explained with the concept of correlation between samples of an experiment (sample set). The variance of the sample mean using systematic sampling is given by [19, 20]

$$Var(\hat{X}_{sys,n}) = \frac{\sigma^2}{n}[1 + (n-1)\rho] \tag{3.40}$$

where $\rho$ is a measure of the correlation between pairs of samples within the same systematic sample.

$$\rho = \frac{E(X_{ij} - \mu)(X_{ij'} - \mu)}{E(X_{ij} - \mu)} \tag{3.41}$$

where, $-\frac{1}{(n-1)} \leq \rho \leq 1$, $i = 1, \ldots, k$ and $j, j' = 1, \ldots, n$, $j \neq j'$.

Thus, a theoretical accuracy of systematic sampling is not practically assessable, as knowledge of all $k$ systematic samples is necessary to calculate the variance of systematic samples. Eq. (3.40) also shows that when $\rho$ is positive, the estimator is *not consistent*, since the accuracy is not increased with large $n$. If $\rho$ is close to 1, then the variability of elements within the sample set is too small compared to the variability among possible sample sets, and systematic sampling will yield a larger variance than using simple random sampling. If $\rho$ is negative, then systematic sampling may be better than simple random sampling. The correlation may be negative if the variability of elements *within* a systematic sample set tends to be larger than *among* systematic sample sets. For $\rho$ close to 0, systematic sampling is roughly equivalent to simple random sampling. When the population is *randomly ordered*, systematic sampling will give us precision approximately equivalent to that obtainable by simple random sampling.

Figure 3.12 illustrates extreme performance of systematic sampling from the same population. Suppose one tries to estimate a population mean with a systematic sampling. For the periodic population shown in Figure 3.12, when $k(=m/n)$ is the same as the period, the value of the sample is the same for all samples in any possible set of samples; thus, an increase of sample size would not increase the accuracy at all. Meanwhile, with $k = 4 (> 3)$, it always gives the exact population mean with a smaller number of samples that is better than random sampling. Therefore, randomness in samples is important to *assess* the accuracy, and to avoid extreme performance.

Scalability of measurement using adaptive random sampling may be further enhanced by infrequent random number generation. Suppose $n$ packets out of $m$ are collected (refer to illustration in Figure 3.13). Rather than generating a random number for each packet (first row), we maintain a counter initialized to $k(=m/n)$. The counter is decremented upon each packet arrival, and when it reaches 0 it is reset back to $k$. Whenever the counter is set to $k$, a random number $i$ (from 0 to $k$) is generated and the $i$th packet, counting from the time of the counter reset, is sampled as illustrated in the second row of the figure. As a further enhancement, a hybrid approach could be used as shown in the third row of Figure 3.13. By changing the starting point randomly several times $(c(< n))$, this process has the effect of shuffling the elements of the population.

## 3.5 Experimental Results

In this section, we first validate our theoretical result with synthetic data. We then empirically evaluate the performance of our adaptive packet sampling technique for flow measurement using real network traces.

In order to verify our theoretical results, first we conduct experiments with synthesized data where all flows are elephants whose proportions are the same as the threshold. In synthesized data, all flows have the same packet count and their durations fall within a block. They have different byte counts caused by various means and standard deviations of their packet sizes. The SCVs for all flows, however, are the same $(SCV = 1.3)$. Two types of traffic data are generated according to the flow's

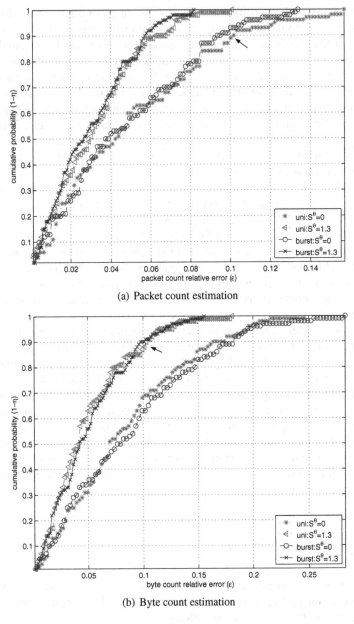

(a) Packet count estimation

(b) Byte count estimation

**Fig. 3.14** Synthetic data: Extremely bursty and uniformly distributed flows with variable packet sizes ($\{\eta, \varepsilon\} = \{0.1, 0.1\}$, $\{p^\theta, S^f\} = \{0.01, 1.3\}$, $p^f = p^\theta$).

packet rate variability. Packets of flows in the first traffic data are uniformly distributed, while flows in the second traffic are bursty. We first use the sampling rate ignoring the packet size variability ($S^\theta = 0$), and repeat the measurements with a higher sampling rate using the actual value of packet size SCV.

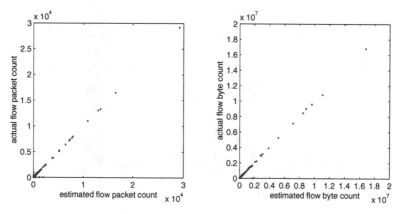

**Fig. 3.15** Actual vs. estimated flow volume (trace $\Pi_1$, $\{\eta, \varepsilon\} = \{0.1, 0.1\}$).

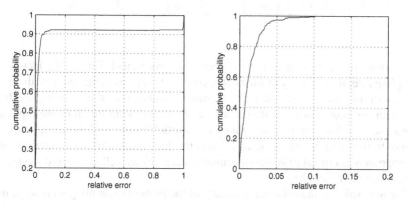

**Fig. 3.16** Relative error of elephant flows (trace $\Pi_1$, $\{\eta, \varepsilon\} = \{0.1, 0.1\}$).

   In both cases, we calculate the cumulative probability of relative errors for packet and byte count estimates. Figure 3.14(a) shows that packet count estimation indeed conforms to the pre-specified accuracy parameter, i.e., the probability of relative errors larger than $\varepsilon = 0.1$ is around $\eta = 0.1$ with $S^\theta = 0$. Meanwhile, as shown in Figure 3.14(b), the probability of relative errors larger than $\varepsilon = 0.1$ is a lot higher than $\eta = 0.1$, when the packet size variability is not considered for the sampling rate. When taking packet size variability into account ($S^\theta = 1.3$), the byte count

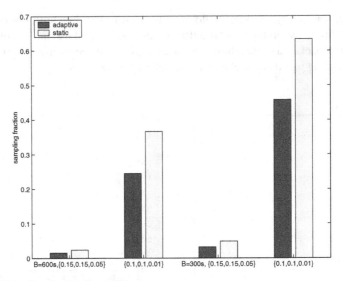

**Fig. 3.17** Sampling fraction (trace $\Pi_1$).

**Table 3.3** Flow cache size reduction.

| Parameters | Trace $\Pi_5$ (24hr) | | Trace $\Pi_1$ (4hr) | | Trace $\Pi_6$ (90sec) | |
|---|---|---|---|---|---|---|
| $\{\eta, \varepsilon, p^\theta, S^\theta\}$ | Adapt. | Static | Adapt. | Static | Adapt. | Static |
| $\{.1, .1, .001, .2\}$ | .459 | .711 | .721 | .772 | .742 | .746 |
| $\{.15, .15, .005, .0\}$ | .433 | .697 | .710 | .735 | .727 | .730 |

estimation error conforms to the pre-specified error bound. Due to the increased sampling rate, the accuracy of the packet count estimates is increased, i.e., the proportion of all estimates whose relative error is smaller than $\varepsilon$ is even higher than the predicted probability $1 - \eta$. The above observations are true for both types of flows and therefore, are independent of the flow's burstiness. Considering fluctuation of traffic, in particular, a high day/night traffic ratio, adapting the sampling rate appropriately is critical to achieve estimation accuracy as well as to avoid too much unnecessary oversampling.

We also validate the achieved accuracy of the proposed sampling technique, using real traces. In Figure 3.15, all the estimated flow volumes using sampled packets are compared to the actual flow volumes, to show the performance qualitatively. It can be observed that small volume flow estimations (indirectly, low proportion flows) are more farther from the actual volumes, while high volume flow estimations (elephants) tend to be closer to the actual volumes. Figure 3.16 shows the cumulative probability of the relative error estimating elephant flows. There are a few flows shortened due to timeout. Their relative errors are close to 1, because 1 or 2 packet flows from the broken flows are compared to the original flow (left plot). However, after removing statistics of those flows (right plot), the cumulative probability of relative error, being less than $\varepsilon$, is higher than $1 - \eta$. Recall that the sampling rate was aimed for an accuracy of the minimum elephant whose proportion is equal to the threshold. For elephants whose threshold is higher than the threshold, the achieved accuracy is supposed to be better. Thus, the statistical accuracy among *all* elephants becomes better than the specified, as in Figure 3.16.

Next, we examine the efficiency of the adaptive sampling in terms of reduction in packet measurement and flow cache size. Note that the amount of measurement in the adaptive scheme depends on the accuracy parameters and elephant flow thresholds, and does not depend on the total traffic. In static sampling, however, the amount of samples is proportional to the total traffic. A sampling fraction, which is the ratio of the number of samples over the total number of packets, determines the resource usage efficiency. We compare the sampling fraction from a trace using adaptive sampling and static sampling. The average sampling rate of the adaptive sampling method is used for the sampling rate of the static sampling method. In adaptive sampling, higher accuracy requires a larger number of samples, while a larger block size decreases the sampling rate. As shown in Figure 3.17, the sampling fraction is higher for the static sampling scheme for various block sizes and accuracy parameters. Since the processing overhead is a function of the number of packets sampled, the advantage of the adaptive sampling is clear.

Reduction in flow cache size is another benefit from sampling, because some flows can be omitted in keeping flow statistics. Table 3.3 shows flow cache size reduction for both adaptive and static sampling with various traces and various sampling rates. Flow cache size reduction is computed in terms of average flow cache size. The case of using sampling is compared to the one without sampling for the trace. Packet sampling indeed reduces flow cache size for all cases of sampling rate and traces. The adaptive sampling performs better than static sampling, particularly when day/night traffic fluctuation is considered. However, its reduction is less relevant to the sampling parameters. This is because as the sampling rate decreases, the timeout value in case of sampling increases accordingly. Then, flow statistics should be kept for a longer time when sampling is used.

## 3.6 Summary

In this chapter, we have addressed the problem of flow volume measurement using the packet sampling approach. Since a small percentage of flows are observed to account for a large percentage of the total traffic, we focused on the accurate measurement of *elephant* flows. We proposed an adaptive sampling method that adjusts the sampling rate so as to bound the error in flow volume estimation without excessive oversampling. The proposed method based on stratified random sampling divides time into strata, and within each stratum, samples packets randomly at a rate determined according to the minimum number of samples needed to achieve the desired accuracy. Unlike static sampling where the amount of samples increases proportionally with the amount of traffic, our scheme collects a *fixed* number of samples on each stratum, regardless of total traffic. Thus, the proposed scheme becomes truely scalable as opposed to static sampling. The technique can be applied to any granularity of flow definition. Through analysis and experimentation we have shown that the proposed method provides an accurate and unbiased estimation of byte and packet counts of elephant flows without excessive oversampling. We have also discussed practical issues and argued that our method can be implemented efficiently to match the line speeds. We conclude that the ability to control the accuracy of estimation, and thus the tradeoff of the utility and overhead of measurement, makes our adaptive sampling method a scalable and attractive solution for flow volume measurement.

# References

1. D. Berry and B. Lindgren. *Statistics theory and Methods*. Duxbury Press, ITP, 2nd edition, 1996.
2. S. Bhattacharyya, C. Diot, J. Jetcheva, and N. Taft. Pop-level and access-link-level traffic dynamics in a tier-1 pop. In *Proceedings of ACM SIGCOMM Internet Measurement Workshop*, San Francisco, November 2001.
3. P. Billingsley. *Probability Measures*. Wiley-Interscience Publication, 1995.
4. M. Borella. Source models of network game traffic. *Computer Communications*, 23(4):403–410, Feb. 2000.
5. CAIDA. The Cooperative Association for Internet Data Analysis. http://www.caida.org.
6. B. Carpenter. Middleboxes: Taxonomy and issues. Internet Engineering Task Force Request for Comments: 3234, February 2002.
7. K. Claffy, H.-W. Braun, and G. C. Polyzos. A parameterizable methodology for internet traffic flow profiling. *IEEE Journal of Selected Area in Communication*, (13):1481–1494, 1995.
8. N. Duffield, C. Lund, and M. Thorup. Charging from sampled network usage. In *Proceedings of ACM SIGCOMM Internet Measurement Workshop*, 2001.
9. A. Erramilli, O. Narayan, and W. Willinger. Experimental queuing analysis with long-range dependent packet traffic. *IEEE/ACM Transactions on Networking*, 4(2):209–223, April 1996.
10. C. Estan and G. Varghese. New directions in traffic measurement and accounting. In *Proceedings of ACM SIGCOMM*, 2002.
11. W. Fang and L. Peterson. Inter-as traffic patterns and their implications. In *Proceedings of IEEE Globecom*, Rio, Brazil, December 1999.
12. A. Feldmann, A. Greenberg, N. Reingold C. Lund, J. Rexford, and F. True. Deriving traffic demands for operational ip networks: Methodology and experience. *IEEE/ACM Transactions on Networking*, pages 265–279, June 2001.
13. J. Gottman. *Time-series analysis*. Cambridge University Press, 1981.
14. IPFIX. Internet Engineering Task Force, IP Flow Information Export. *http://www.ietf.org/html.charters/ipfix-charter.html*.
15. Juniper packet sampling. http://www.juniper.net/techpubs/software/junos53/swconfig53-policy/html/sampling-overview.html.
16. PSAMP. Internet Engineering Task Force Packet Sampling working group. *https://ops.ietf.org/lists/psamp*.
17. B. Ryu, D. Cheney, and H. Braun. Internet flow characterization: Adaptive timeout strategy and statistical modeling. In *Proceedings of Passive and Active Measurement Workshop*, Amsterdam, April 2001.

18. Cisco sampled netflow.    http://www.cisco.com/univercd/cc/tc/doc/product/software/ios120/
    120newft.
19. T R. Scheaffer, W. Mendenhall, and R. Ott. *Elementary Survey Sampling*. Duxbury Press, 5th
    edition, 1995.
20. T. Yamane. *Elementary Sampling Theory*. Prentice-Hall, Inc., 1967.

# Part II
# Scalable Delay Monitoring with Active Probing

# Chapter 4
# Analysis of Point-To-Point Delay in an Operational Backbone

**Abstract** In this chapter, we perform a detailed analysis of point-to-point packet delay in an operational tier-1 network. The point-to-point delay is the time between a packet entering a router in one PoP (an ingress point) and its leaving a router in another PoP (an egress point). It measures the one-way delay experienced by packets from an ingress point to an egress point across a carrier network and provides the most basic information regarding the delay performance of the carrier network. Using packet traces captured in the operational network, we obtain precise point-to-point packet delay measurements and analyze the various factors affecting them. Through a simple, step-by-step, systematic methodology and careful data analysis, we identify the major network factors that contribute to point-to-point packet delay and characterize their effect on the network delay performance. Our findings are: 1) delay distributions vary greatly in shape, depending on the path and link utilization; 2) after constant factors dependent only on the path and packet size are removed, the 99th percentile variable delay remains under 1 ms over several hops and under link utilization below 90% on a bottleneck; 3) a very small number of packets experience very large delay in short bursts.

**Key words:** point-to-point packet delay, delay measurement, backbone network, delay distributions, equal cost multiple path, queueing delay

## 4.1 Introduction

Network delay is often a key performance parameter in Service Level Agreements (SLAs) that specify the network performance service targets agreed upon between an Internet service provider (ISP) and its customer [12]. In most of today's SLAs, an average packet delay over a relatively long period of time is usually specified. Such an average delay value is typically estimated using crude measurement tools, such as ping. It is generally not quite representative nor indicative of the performance a specific customer will experience, but simply a very rough estimate of the overall delay performance of an ISP network [6]. Given the increasing critical nature of network performance to many e-business operations and the competitive nature of the ISP market, more precise specification of network performance requirements (such as packet delay) in SLAs will become a differentiating factor in ISP service offerings. Hence, a careful analysis of packet delay performance in an operational network is imperative.

In this chapter, we carry out a large-scale delay measurement study using packet traces captured from an operational tier-1 carrier network. We focus on the so-called *point-to-point* (or router-to-router) delay – the time between a packet entering a router in one PoP (the ingress point) and its leaving a router in another PoP (the egress point). Previously, Papagiannaki et al. [20] measured and analyzed single-hop delay in a backbone network. Our work is a natural extension of its work onto multiple hops. The point-to-point delay measures the one-way delay experienced by packets from an ingress point to an egress point across a carrier network, and provides the most basic information regarding the delay performance of the carrier network. The objective of our study is two fold: 1) to analyze and characterize the point-to-point packet delays in an operational network; and 2) to understand the various factors that contribute to point-to-point delays and examining the effect they have on the network delay performance.

Delay between two end-users (or points) has been studied extensively for its variation, path symmetry, queueing delay, correlation with loss, and more in [3, 22]. Because of its direct implication on delay-sensitive applications, such as VoIP (Voice over IP) and video streaming, and user-perceived performance of web downloads, there are continuing efforts on measuring, monitoring, and analyzing the end-to-end delay. Since the end-to-end delay is over several hops and may reflect route changes, it is not easy to pinpoint a cause of significant change, when we observe one. Point-to-point delay in an ISP is a building block of end-to-end delay, and understanding the main factors of point-to-point delay will add insight to the end-to-end delay.

The study of the point-to-point packet delay poses several challenges.

- In order to understand the evolution of point-to-point delay, packet measurements need to span over a long period of time (e.g. hours).
- Data should be collected simultaneously from at least two points within an ISP, and clocks should be globally synchronized to compute one-way delay.
- Routing information is needed to keep track of route changes, if any. Other supplementary data, such as fiber maps and router configuration information, is needed to address path-specific concerns.

In our monitoring infrastructure, we have addressed all of the above points [10], and we believe this is a first study that focuses on point-to-point delay within an ISP.

We use packet traces captured in an operational tier-1 ISP network. Packets are passively monitored [10] at multiple points (routers) in the network with GPS-synchronized timestamps. Captured packets between a pair of any two points are the entire traffic between the two, not from active probes or from a certain application. Using these precise delay measurement data, as well as SNMP (Simple Network Management Protocol) [28], router configurations, routing updates, and fiber maps, we perform a careful analysis of the point-to-point delay distributions and develop a systematic methodology to identify and characterize the major network factors that affect the point-to-point packet delays.

Our observations and findings are the following. First, the point-to-point packet delay distributions, in general, exhibit drastically different shapes, often with multiple modes, that cannot be characterized by a single commonly known mathematical distribution (e.g., normal or heavy-tailed distribution). There are many factors that contribute to these different shapes and modes. One major factor is the equal-cost multi-path (ECMP) routing [29] commonly employed in operational networks. It introduces multiple modes in point-to-point delay distributions. Another major factor is the packet size distribution that has a more discernible impact on point-to-point packet delay distributions when the network utilization is relatively low. By identifying and isolating these factors through a systematic methodology with careful data analysis, we then focus on the variable delay components and investigate the role of link utilization in influencing the point-to-point packet delays. We find that when there is no bottleneck of link utilization over 90% on a path, the 99th percentile variable delay is less than 1 ms. When a link on the path has link utilization above 90%, the weight of the variable delay distribution shifts and the 99th percentile reaches over 1 ms. Even when the link utilization is below 90%, a small number of packets experience delays of an order of magnitude larger than the 99th percentile, and affect the tail of the distribution beyond the 99th percentile.

In summary, the contribution of this work first lies in the detailed analysis and characterization of the point-to-point packet delays and their major components in an operational network. We also develop a systematic methodology to identify, isolate, and study the main contributing factors. Understanding when and how they affect the point-to-point packet delays in an operational network is important both theoretically and in practice: such an understanding will not only help network engineers and operators to better engineer, operate, and provision their networks, it also provides valuable insight as to the important performance issues in operational networks that network researchers should pay attention to. In particular, our study sheds light on the necessity and importance of devising more representative and relevant measurement mechanisms for network delay performance and link utilization monitoring.

The remainder of the chapter is structured as follows. In Section 4.2, we discuss available techniques and tools to measure network performance. Section 4.3 lays out the factors that affect the point-to-point delay and describes our packet measurement methodology and other data, such as SNMP statistics and router configuration information. The main discussion begins in Section 4.5 with general observations on point-to-point delay. Then in Section 4.6 we isolate constant factors that are fixed on the path and the packet size. In Section 4.7, we focus only on the variable delay that is due to cross traffic. In Section 4.8, we investigate further SNMP data for the cause of long delays. In Section 4.9, we summarize our findings and discuss their implications on current monitoring practices.

## 4.2 Related Works

In this chapter, we summarize measurement infrastructure as well as available tools and techniques to measure network performance. The references and links of the above tools can be found at the repository of CAIDA [4]. Most performance measurement tools conduct in an active manner rather than in a passive measurement.

First, we summarize public performance measurement infrastructures.

*Internet2 (I2) Abilene* [1] is an advanced backbone network that connects regional network aggregation points, called gigaPoPs, to support the work of Internet2 universities as they develop advanced Internet applications. The Abilene Project complements other high-performance research networks. Abilene enables the testing of advanced network capabilities prior to their introduction into the application development network. These services are expected to include Quality of Service (QoS) standards, multicasting, and advanced security and authentication protocols.

*SLAC/DOE/ESnet* [13], High Energy and Nuclear Physics use pingER tools on 31 monitoring sites to monitor network performance for over 3000 links in 72 countries. Monitoring includes many major national networks (including ESNet, vBNS, and Internet2-Abilene), as well as networks in South America, Canada, Europe, the former Soviet Union, New Zealand, and Africa.

The *Measurement and Analysis of Wide-area Internet (MAWI) Working Group* [19] studies the performance of networks and networking protocols in Japanese wide-area networks. Sponsored by the Widely Integrated Distributed Environment (WIDE) project, MAWI is a joint effort of Japanese network research and academic institutions with corporate sponsorship. Mantra *Monitor and Analysis of Traffic in Multicast Routers (Mantra)* [18] monitors various aspects of global Internet multicast behavior at the router level. Visualization snapshots and accompanying tables are updated every 15-30 minutes.

*NIMI* is a project, begun by the National Science Foundation and currently funded by DARPA, to measure the global Internet [23]. Based on Vern Paxson's Network Probe Daemon, NIMI was designed to be scalable and dynamic. NIMI is dynamic in that the measurement tools employed are treated as third party packages that can be added or removed as needed. For example, the MINC (Multicast Inference of Network Characteristics) measurement methodology for determinng performance characteristics of the interior of a network from edge measurements has been tested and validated using the NIMI infrastructure.

*CAIDA's Measurement and Operations Analysis Team (MOAT)* [4] has created a Network Analysis Infrastructure (NAI) to derive a better understanding of system service models and metrics of the Internet. This includes passive measurements based on analysis of packet header traces (link to PMA above); active measurements (link to AMP above); SNMP information from participating servers; and Internet routing related information based on BGP data.

*RIPE (Reseaux IP Europeens)* [25] is a collaborative organization open to organizations and individuals, operating wide area IP networks in Europe and beyond. The objective of RIPE is to ensure the administrative and technical coordination necessary to enable operation of a pan-European IP network. RIPE does not operate a network of its own. Currently, more than 1000 organizations participate in the work. The result of the RIPE coordination effort is that an individual end-user is presented with a uniform IP service on his or her desktop irrespective of the particular network his or her workstation is attached to.

*Surveyor* [17] is a measurement infrastructure that is being currently deployed at participating sites around the world. Based on standard work done in the IETF's IPPM Working Group, Surveyor measures the performance of the Internet paths among participating organizations. The project is also developing methodologies and tools to analyze the performance data.

The *Canadian national research facility* uses perl scripts to trace paths toward nodes of interest to *TRIUMF* [30]. Packet loss summaries and graphs are generated daily from pins made at 10 minute intervals. Traceroute data is gathered four times daily. Network visualization maps are generated from the traceroute data.

The *Route Views router*, `route-views.oregon-ix.net`, [26] is a collaborative endeavor to obtain real-time information about the global routing system from the perspectives of several different backbones and locations around the Internet. It uses multi-hop BGP peering sessions with backbones at interesting locations (note that location should not matter if the provider is announcing consistent routes corresponding to its policy). Route Views uses AS65534 in its peering sessions, and routes received from neighbors are never passed on nor used to forward traffic. Finally, route-views.oregon-ix.net itself does not announce any prefixes.

The *WAND (Waikato Applied Network Dynamics)* project [31] aims to build models of Internet traffic for statistical analysis and for the construction of simulation models. The project builds its own measurement hardware and collects and archives significant network traces. These are used internally and are also made available to the Internet research community. Traces are accurately timestamped and synchronized to GPS. Many traces are 24 hours long, some are up to a week long, and there are plans to provide even longer traces in the future. The WAND project is based at the University of Waikato in New Zealand with strong collaboration from the University of Auckland.

Some tools are designed to measure more than one performance metric, and most of performance measurement tools employ active probing techniques. Here, we briefly describe the tools in an alphabetical order.

`bing` [2] is a point-to-point bandwidth measurement tool based on ping. `bing` determines the raw - as opposed to available or average - throughput on a link by measuring ICMP echo requests round trip times for different packet sizes for each end of the link. It deduces the round trip time of the link of interest, and then, the raw capacity of the link is computed exploiting the round trip time of different packet sizes.

*CAIDA's skitter tool* is used to visualize topology and performance attributes of a large cross-section of the Internet by probing the path from a few sources to many thousands of destinations spread throughout the IPv4 address space.

`clink` [9] reimplementes `pathchar` to generate interval estimates for bandwidth, using the so called even-odd technique [9]. When it encounters a routing instability, it collects data for all the paths that are met, until one of the paths generates enough data to yield an estimate.

iperf [14] is a tool for measuring maximum bandwidth. It allows the tuning of various parameters and UDP characteristics. It reports bandwidth, delay jitter, and datagram loss.

netperf is a benchmark that can be used to measure the performance of many different types of networking. It provides tests for both unidirectional throughput, and end-to-end latency.

pathchar [15] estimates performance characteristics of each node along a path from a source to destination. It leverages the ICMP protocol's Time Exceeded response to packets whose TTL has expired. Sending a series of UDP packets of various sizes to each hop, pathchar uses knowledge about earlier hops and the round trip time distribution to this hop to assess incremental bandwidth, latency, loss, and queue characteristics across this link.

pathload [16] estimates a range of available bandwidths for an end-to-end path. The available bandwidth is the maximum IP-layer throughput that a flow can get in the path from sender to receiver, without reducing the rate of the rest of the traffic in the path. pathload consists of processes running at the sender and receiver, respectively. The sender sends periodic streams of UDP packets to the receiver at a certain rate. Upon the receipt of a complete fleet, the receiver checks if there is an increasing trend in the relative one-way packet delays in each stream. The increasing trend indicates that the stream rate is larger than the available bandwidth.

pathrate [8] estimates the capacity of Internet paths, even when those paths are significantly loaded. pathrate is based on the dispersion of packet pairs and packet trains. First, many packet pairs of variable sizes are sent from the source of the path to the sink in order to derive a set of local modes. One of these local modes is the capacity of the path. Then, long packet trains are sent and their total dispersion is measured. The asymptotic dispersion rate (ADR) is calculated from the dispersion of these long packet trains. Given that the capacity of the path is larger than the ADR, any local modes less than the ADR are rejected. From the modes that remain, the strongest and narrowest local mode is selected as the capacity estimate.

sprobe measures bottleneck bandwidths in both directions of a network path with TCP SYN/RST packets. sprobe uses the packet pair technique and exploits properties of the TCP protocol in a similar manner used in [27].

treno is an Internet TCP throughput measurement tool based on a user-level implementation of a TCP-like protocol. This allows it to measure throughput independently of the TCP implementation of end hosts and to serve as a useful platform for prototyping TCP changes.

## 4.3 Point-to-Point Delay: Definition and Decomposition

We define *point-to-pont* packet delay as the time between a packet entering a router in one PoP (the ingress point) and its leaving a router in another PoP (the egress point). Theoretically speaking, we can decompose the point-to-point packet delay into three components: propagation delay, transmission delay and queueing delay. Propagation delay is determined by physical characteristics of the path a packet traverses, such as the physical medium and its length. Transmission delay is a function of the link capacities along the path, as well as the packet size. Queueing delay depends on the traffic load along the path, and thus varies over time. In practice, many other factors can contribute to the delay packets experience in an operational network. First, network routing may change over time; hence, the path between an ingress point and an egress point may not be fixed. Furthermore, in today's high-speed carrier networks, equal-cost multi-path (ECMP) routing is commonly employed for traffic engineering and load balancing. Thus, packets going through the same ingress-egress point pair may take different paths with differing path characteristics such as propagation delay, link capacities, and traffic load. These factors can introduce significant variations in point-to-point delay measurement. We will refer to factors that depend solely on path characteristics as well as packet sizes as constant factors, since these factors have a fixed effect on point-to-point delays experienced by packets of the same size that traverse exactly the same path.

$$d := d^{fixed} + d^{var} \tag{4.1}$$

Queuing delay introduces a *variable* component to the point-to-point delays experienced by packets, as it depends on the traffic load (or cross traffic) along the path that varies at different links and with time. In addition to traffic load, sometimes anomalous behaviors of individual routers or the network can also introduce other unpredictable factors that affect point-to-point delays packets experience. For example, in an earlier study [20], authors discover that packets may occasionally experience very large delays. Such large delays can be caused by a router performing other functions such as routing and forwarding table updates, etc. In addition, during the routing protocol convergence, transient forwarding loops [11] may occur, and packets caught in such loops will suffer unusually long delays. These very large delays are obviously not representative of the typical delay performance of a network, and thus should be considered as outliers in delay measurement.

## 4.4 Data Collection and Delay Measurement Methodology for Analysis

With a basic understanding of the various factors affecting point-to-point packet delays in an operational network, we now proceed to describe the delay measurement setting and methodology employed in our study. The packet traces we use in this study are from the Sprint IPMON project [10]. On the Sprint IP backbone network, about 60 monitoring systems are deployed at 4 PoPs (Point-of-Presences), each with many access routers and backbone routers. Using optical splitters, the monitoring systems capture the first 44 bytes of all IP packets and timestamp each

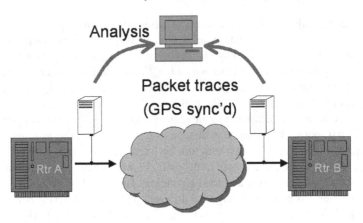

**Fig. 4.1** IPMon data collection for delay.

of them. As the monitoring systems use the GPS (Global Positioning System) for their clock synchronization, the error in timestamp accuracy is bounded to less than 5 $\mu$s.

To obtain packet delays between two points (i.e., between two routers in two different PoPs), we first identify those packets that traverse those two points of measurement. We use hashing to match packets efficiently. Only 30 bytes out of the first 44 bytes of a packet carry distinguishing information and are used in hashing. IP header fields, such as the version, TTL (Time-To-Live), and TOS (Type of Service), are not used. For more detail in packet matching, refer to [20].

In performing packet matching, we occasionally find duplicate packets in our traces. They are likely due to unnecessary link-level retransmissions or routing loops, and have been reported in [11, 20, 21]. Since duplicate packets have the identical 30 bytes we use in hashing, we cannot always disambiguate matching packets in the corresponding traces and do not use them in our analysis. In all pairs of the traces, we observe that duplicate packets are less than 0.05% of the total number of packets.

We have analyzed weeks' worth of concurrent traces dispersed over a few years. For this work, we select traces from 2 dates: August 6th, 2002 and November 21st, 2002. The main criterion used in trace selection is the number of packets common in a pair of traces. Not all pairs of traces have many packets in common. Traffic entering a network at one ingress point is likely to split and exit through many egress points. Thus, the number of packets observed on a pair of monitored links depends on the actual amount of traffic that flows from one to the other. Some pairs of traces have no packet in common. As our goal is to analyze packet delay, we choose to use pairs of traces with the most matches in our analysis.

In conducting our delay measurement study, we have analyzed all matched trace pairs from August 6th, 2002 and November 21st, 2002. However, for concise illustration, we choose only three pairs as representative examples to illustrate our observations and findings, as shown in Table 4.1. The delay statistics of the three

**Table 4.1** Summary of Matched Traces: Delays are in milliseconds.

| Data Set | From | To | Start Time (UTC) | Duration | Packets |
|---|---|---|---|---|---|
| 1 | OC-48 | OC-12 | Aug. 6, 2002 12:00 | 16h 24m | 1,349,187 |
| 2 | OC-12 | OC-12 | Nov. 21, 2002 14:00 | 3h | 498,865 |
| 3 | OC-12 | OC-48 | Nov. 21, 2002 14:00 | 5h 21m | 4,295,781 |

**Table 4.2** Delay statistics of Matched Traces: Delays are in milliseconds.

| Data Set | Min | Mean | Median | 99th | Max. |
|---|---|---|---|---|---|
| 1 | 28.432 | 28.458 | 28.453 | 28.494 | 85.230 |
| 2 | 27.949 | 28.057 | 28.051 | 28.199 | 55.595 |
| 3 | 28.424 | 31.826 | 32.422 | 34.894 | 100.580 |

trace sets we use in this chapter are shown in Table 4.2. In all three sets, the source traces are from the West Coast and the destination traces are from the East Coast. The path between a source and a destination consists of more than 2 hops, and thus the delay reported in this work is over multiple hops. The duration of matched traces varies from around 4 hours to more than 16 hours.

In addition to the packet traces, we also use other network data such as SNMP statistics on link and CPU utilization, router configuration files, and fiber maps in our analysis and identification of various network factors that affect the measured point-to-point packet delays. Using the router configuration files, we obtain the information about the paths, associated IP links, and router interfaces that packets traverse in the point-to-point packet delay measurements. The fiber map provides us with the further information about the fiber links and estimated propagation delay of the paths. SNMP data that reports the link load and router CPU utilization averaged over five-minute intervals, is also collected on every link of the paths, and are used to correlate the point-to-point packet delay measurements and 5-minute average network utilization.

## 4.5 General Observations on Delay Distributions

We begin this section with general observations on point-to-point packet delay distributions obtained from the three trace sets in Table 4.1. First, we note from Table 4.2 that the minimum delays from all three delay distributions are about 28 ms, which reflect the transcontinental delay of the U.S. Other statistics such as mean, median, 99th percentile,[1] and maximum delays show more variations. However, when we examine the delay distributions of the three sets, the difference between the traces are striking. We use 1 $\mu$s bins to plot the empirical density distribution

---

[1] Let $F(x)$ be a cumulative distribution function of a random variable $x$. $q = F^-(0.99)$ is the 99th percentile of the distribution of $x$.

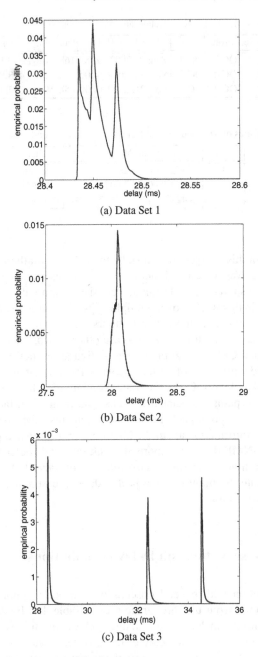

Fig. 4.2 Point-to-point packet delay distributions.

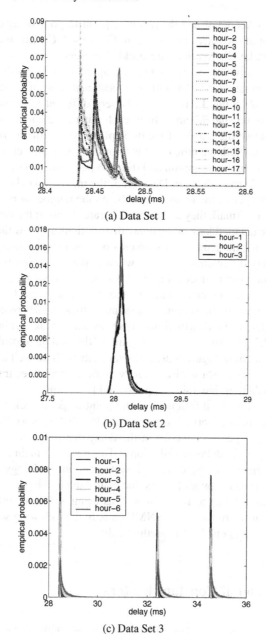

(a) Data Set 1

(b) Data Set 2

(c) Data Set 3

**Fig. 4.3** Hourly point-to-point packet delay distributions.

functions of the packet delay, and the resulting point-to-point delay distributions in the entire duration of the traces are shown in Figure 4.2. Clearly, the shapes of the three (empirical) delay distributions are starkly different.

Figure 4.2(a) exhibits three peaks that are apart from each other only in tens of microseconds. Figure 4.2(b) has only one peak and most of the distribution lies between 27.9 and 28.5 ms. Figure 4.2(c) is very different from the other two: it has three unique peaks that are apart from each other by 2 and 4 ms, respectively. Here we point out that the x-axes of the three plots are in different ranges: they are chosen to include the 99th quantile point but not the maximum delay point. Though the difference between the minimum and the 99th percentile delay is less than 1 ms in Data Sets 1 and 2, and 6.5 ms in Data Set 3, the maximum delay is significantly larger than the 99th percentile in all three sets. As the number of packets with such extreme delay is very small, they represent very rare events in the network.

We take a more detailed look at the delay distributions and how they change over time. We divide each trace pair into segments of an hour long and plot the hourly point-to-point delay distributions and see whether they look significantly different from the that of the overall traces. Figure 4.3 shows the hourly delay distributions overlaid on top of each other for each of the three trace sets. Clearly, the basic shapes of the three distributions remain the same in these hourly plots, in particular, the peaks or modes of the distributions; there are three peaks within a very short range in Figure 4.3(a), a single peak in Figure 4.3(b), and three peaks with a great distance between them in Figure 4.3(c). However, the bulk as well as the tail of the hourly delay distributions show discernible variations in all three trace pairs: some hourly delay distributions have shorter peaks and fatter tails.

What contributes to the differences in the point-to-point packet delay distributions in an operational network? More specifically, what network factors cause the differing numbers of peaks or modes in the delay distributions? What affects the bulk and (or) tail of the delay distributions? These are the main questions we set out to answer in this work. We develop a systematic methodology to identify and characterize the various network factors that contribute to the point-to-point packet delays through careful data analysis, as well as using additional information such as router configuration, routing, and SNMP data. In the following sections, we will investigate these factors one by one methodically.

## 4.6 Identification of Constant Factors

In this section we characterize and isolate the constant network factors that have a fixed effect on point-to-point delays experienced by packets of the same size traversing along the same *physical* (i.e., fiber) path. In particular, we identify and analyze the constant network factors that contribute to the *modes* in the point-to-point packet delay distributions shown in the previous section. We suspect the modes of delay distribution that are spaced with relatively large distances (1 ms or more), such as in Figure 4.2(c), are most likely caused by the equal-cost multi-path (ECMP) routing, commonly employed by today's operational networks. Whereas, the modes that are more closely spaced (within 10s or 100s of microseconds), such as in Figure 4.2(a), are probably due to the effect of various packet sizes that incur different transmission delays. The latter factor, for example, has been shown to have an impact on single-hop packet delays [20]. In the following, we develop a systematic methodology to identify and isolate such effects.

## 4.6.1 Equal-Cost Multi-Path Effect

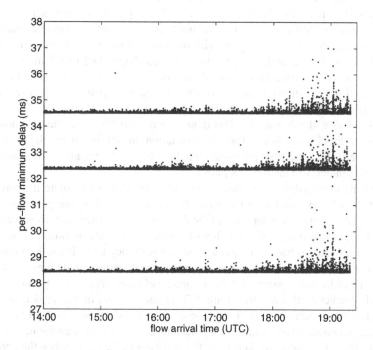

**Fig. 4.4** Minimum flow delay vs. flow arrival time of Data Set 3.

As mentioned earlier, equal-cost multi-path (EMCP) routing is commonly used in today's tier-1 operational networks for load balancing and traffic engineering. Here, equal cost refers to the weights assigned by intra-domain routing protocols such as ISIS. Such weights are not necessarily related to the physical properties such as propagation delay of the paths. In fact, sometimes paths that follow separate fiber circuits are preferred for fault tolerance. Because of the differing characteristics of these physical paths, their propagation and transmission delays may also be different. In using EMCP routing, routers (e.g., Cisco routers in our study) randomly split traffic using a hash function that takes the source IP address, the destination IP address, and the router ID as input and determines the outgoing link for each packet[2]. Therefore packets with the same source and destination IP addresses always follow the same path.

To identify the effect of ECMP routing on point-to-point packet delays, we employ the following method. We first define a *(two-tuple) flow* to be a set of packets with the same source and destination IP addresses, and group packets into flows. We then compute the *minimum* packet delay for each flow. The idea is that by considering the minimum packet delay of a flow, we attempt to minimize the variable delay components such as queueing and retain mostly the fixed delay components. If the minimum delays of two flows differ significantly, they are more likely to follow two different paths. In Figure 4.4 we plot the minimum delay of each flow by the arrival time of the first packet in the flow for Data Set 3. The plot demonstrates the presence of three different paths, each corresponding to one peak in the delay distribution of Figure 4.2(c). We number the path with the smallest delay (28.4 ms), Path 1, the next (32.4 ms), Path 2, and the last with the largest delay (34.5 ms), Path 3.

We use other network data to corroborate our findings. Using the router configuration and fiber path information, we identify the three paths and the exact links the packets in Data Set 3 traverse. The topology map is in Figure 4.6. The source is marked link s, the destination, link d, and the three paths share one common link (denoted as link B), one hop before the destination. In addition, using the fiber map, we also verify that the fiber propagation delay (28 ms, 32 ms, and 34 ms) matches the minimum delay of each of the paths.

With the knowledge of the three paths, the next task is to isolate the effect of ECMP, namely, to classify and separate the packets based on the path they take. From Figure 4.4, it is clear for most of the flows which path they take. However, near the end, minimum delays of some flows become close to the minimum of the next path, and it becomes hard to distinguish which path they take. Thus, the minimum delay of a 2-tuple flow alone is not sufficient to pin down the path it takes during this period. To address this issue, we take advantage of other fields in the packet header. Every IP packet carries a 1-byte long TTL (Time-to-Live) in the header. Because the TTL value is decremented by one at every hop, the difference in the TTL values between two points of observation tells the number of hops between them. When we examine the TTL field of packets from the first hour of the trace (when the per-flow minimum delay can easily tell the path a flow took), packets from those flows whose minimum delay is close to 28.4 ms have a TTL delta of 4. That is, Path 1 consists of four hops from the source to the destination. Other flows, whose minimum delay is above 32.4 ms, all have a TTL delta of 5.

Using the TTL information, we separate packets that follow Path 1 from Data Set 3. For the remaining packets, we classify those flows with delay less than 34.5 ms to Path 2, and the rest to Path 3. Because Paths 2 and 3 have the same TTL delta, near the end of the trace we cannot completely disambiguate the paths some flows take

---

[2] By having the router ID as input to the hash function, each router makes a traffic splitting decision independent of upstream routers.

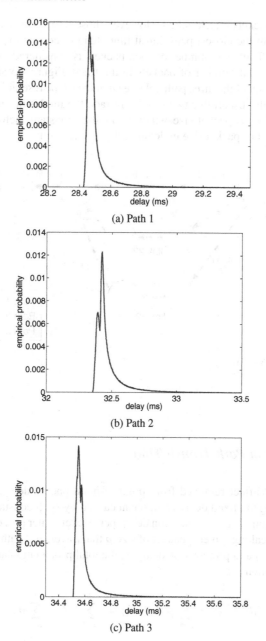

(a) Path 1

(b) Path 2

(c) Path 3

**Fig. 4.5** Delay distributions.

and have some packets misclassified. However, those flows whose minimum delay is high away from the closest path transit time, tend to consist of only a few (one or two) packets. Thus, the number of such packets is considered extremely small, compared to the total number of packets in the trace. Figure 4.5 shows the delay distribution for each of the three paths. We see that the shape resembles Figure 4.2(b) more with a barely discernible two modes. In fact, the modes in Figure 4.2(a) and Figure 4.5 are due to packet size which is another constant factor of delay. We discuss the impact of packet size in detail in the next section.

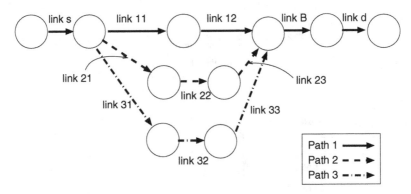

**Fig. 4.6** Path connectivity.

### 4.6.2 Minimum Path Transit Time

With the ECMP effect removed from point-to-point packet delay, we now focus on characterizing the fixed delay component caused by such constant network factors as propagation delay, transmission delay, per-packet router processing time, etc. Theoretically speaking, given a packet of size $p$ that traverses a path of $h$ hops, each link of capacity $C_i$ and propagation delay $\delta_i$, the total propagation and transmission delay can be written as:

$$d^{fixed}(p) = \sum_{i=1}^{h}(p/C_i + \delta_i) = p\sum_{i=1}^{h}1/C_i + \sum_{i=1}^{h}\delta_i.$$

In other words, the fixed delay component (i.e., the total propagation and transmission delay) for a packet of size $p$ is a *linear* (or precisely, an *affine*) function of its size,

$$d^{fixed}(p) = \alpha p + \beta \tag{4.2}$$

where $\alpha = \sum_{i=1}^{h}1/C_i$ and $\beta = \sum_{i=1}^{h}\delta_i$.

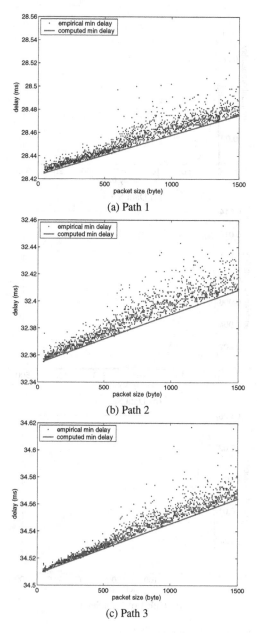

(a) Path 1

(b) Path 2

(c) Path 3

**Fig. 4.7**  Minimum packet delay vs. packet size.

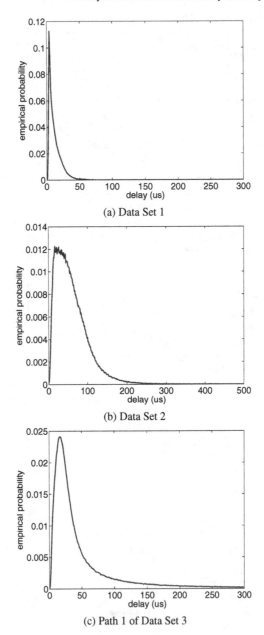

(a) Data Set 1

(b) Data Set 2

(c) Path 1 of Data Set 3

**Fig. 4.8** Delay distributions after removing $d^{fixed}(p)$.

In practice, however, in addition to link propogation and transmission delay, routers also introduce other factors such as per-packet router processing time, backplane transfer inside a router, and so forth that are (mostly) dependent on packet size. Therefore, in reality we believe that such a linear relation between the fixed delay component and packet size is still valid, albeit the parameters $\alpha$ and $\beta$ will not be a simple function of link transmission capacities and propogation delay as in (4.2). In fact, in analyzing a single hop packet delay, the authors in [20] show that the minimum router transit time through a single router is linearly proportional to the packet size. Assuming that the same linear relation holds at all routers on the path, we believe the minimum time any packet experiences on a path of multiple hops is also linearly proportional to the packet size. To validate this assumption, we check the minimum delay of packets of the same size for each path, and plot the minimum delay against the packet size. Figure 4.7 shows the corresponding plots for the three paths in Data Set 3 using packets from the first hour. As expected, there is an apparent linear relation. We fit a line through linear regression on the packet sizes of 40 and 1500, as they are the most common sizes and their minimum delays are most likely to represent the real minimum on the path. This line yields a minimum fixed delay for a given packet size, and we refer to it as the *minimum path transit time* for the given packet size, denoted by $d^{fixed}(p)$. The $d^{fixed}(p)$ was the same with the packets from other hours. Using routing information, we confirmed that the routing was stable for the duration of trace collection.

The parameters $\alpha$ and $\beta$ derived from the plots in Figure 4.7 are listed in Table 4.3. Note in particular that Path 2 and Path 3 have the same $\alpha$ value, but different $\beta$ values. This is consistent with the fact that the two paths have exactly the same number of hops, same link transmission speed, and same type of routers, but slightly different propogation delay along their respective paths due to the difference of their fiber paths. Using packets from other hours in Data Set 3, we obtain almost identical $\alpha$ and $\beta$ values for each path, again confirming the linear relation between the minimum path transit time and packet size. The same result holds for other data sets we have analyzed.

With the fixed delay component $d^{fixed}(p)$ identified, we can now substract it from the point-to-point delay of each packet to study the *variable delay component* (or simply, *variable delay*). Let $d$ represent the point-to-point delay of a packet. The variable delay component of the packet, $d^{var}$, is given by $d^{var} := d - d^{fixed}(p)$. In the next section we investigate in detail the distributions of variable delays, $\{d^{var}\}$, experienced by the packets, how they change over time, and what are the major factors that contribute to long variable delays.

In Figure 4.8 we plot the distribution of variable delay (i.e., after the fixed delay component has been removed) for Data Set 1, Data Set 2, and Path 1 of Data Set 3. The minor peaks we observe in Figures 4.2(a) and Figure 4.5 disappear, and now we only see uni-modal distributions in all the figures[3].

---

[3] Due to space limitation, we do not include the variable delay distributions for Paths 2 and 3 of Data Set 3. They are similar to that of Path 1.

**Table 4.3** Slope and y-intercept of minimum path transit time.

| Data Set | Path | y-intercept | slope |
|----------|------|-------------|-------|
| 1        |      | 28.430931   | 0.00002671 |
| 2        |      | 27.947616   | 0.00005959 |
| 3        | 1    | 28.423489   | 0.00003434 |
| 3        | 2    | 32.353368   | 0.00003709 |
| 3        | 3    | 34.508368   | 0.00003709 |

## 4.7 Analysis of Variable Components

### 4.7.1 Variable Delay and Link Utilization

To understand how the distribution of variable delay changes over time, we plot the hourly distributions of the variable delay for Data Set 1, Data Set 2, and Path 1 of Data Set 3 in Figure 4.9. The hourly distributions are overlaid over each other for ease of comparison. Here we use the complimentary cumulative distribution function (CCDF), as it more clearly illustrates where the bulk of the distribution lies and how the tail tapers off.

From Figure 4.9(a), the hourly distributions of variable delay for Data Set 1 are nearly identical, signifying that the network condition did not change much throughout the entire duration of our measurement. Also, the bulk of the distribution (e.g., 99.9th percentile) lies under 200 $\mu$s. Data Set 1 is from a path over 4 router hops, and thus less than 200 $\mu$s of variable delay is quite small. Hence, packets traversing along this path experience very little queueing delay.

In Figure 4.9(b), the variable delay distributions display a slight shift from the first to the third hour, indicating that the network condition during the 3 hour period has slightly changed. However, the bulk of the distributions (99.9th percentile) is still within hundreds of microseconds. Variable delays of less than 1 ms are generally not very significant, especially over multiple hops, and reflect well on the network performance an end-user should perceive.

The hourly distributions of Path 1 to Path 3 in Data Set 3, however, tell a very different story. The hourly plots shift significantly from hour to hour. The plots from the last two hours diverge from that of the first hour drastically, especially for the tail 10% of the distribution. The 99% delay in the first hour is at 100 $\mu$s, while in the last hour it is at least an order of magnitude larger.

To examine the changes in the variable delay more closely, we zoom in and plot the delay distributions using smaller time intervals. Figure 4.10(a) shows the distribution of variable delay in the first and last three 10-minute intervals for Path 1 of Data Set 1; Figure 4.10(b) shows those from the first and last three 30-minute intervals. In the first 30 minutes, there is little change, and even within the first 90 minutes, there is not much change in variable delay except for the tail beyond the probability of $10^{-3}$. The 99.9th percentile of the distributions is still within a few

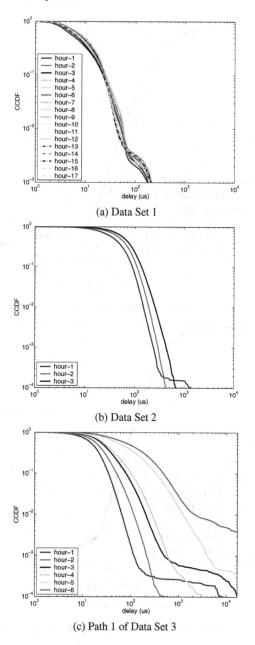

(a) Data Set 1

(b) Data Set 2

(c) Path 1 of Data Set 3

**Fig. 4.9** Hourly distribution of variable delay.

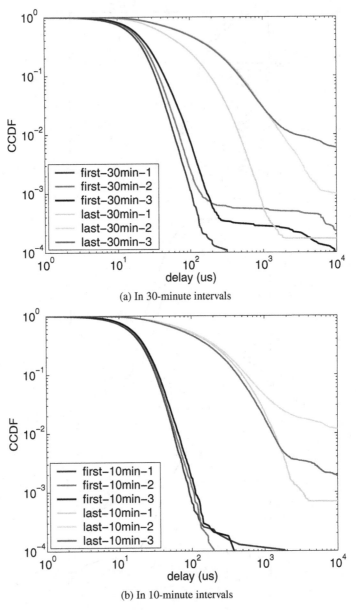

(a) In 30-minute intervals

(b) In 10-minute intervals

**Fig. 4.10** CCDF of $\{d_{px}^-\}$ of the first and last segments of Path 1 of Data Set 3.

100 $\mu$s. However, in the last 90 minutes, in particular, the last 30 minutes, many packets experience a much larger variable delay, causing the 99th percentile delay to reach 2 ms, and in case of one 10-min interval, up to almost 10 ms. Though not presented here due to space limitation, the hourly plots from Paths 2 and 3 exhibit a similar shift toward the end of the trace, and the variable delay increased significantly.

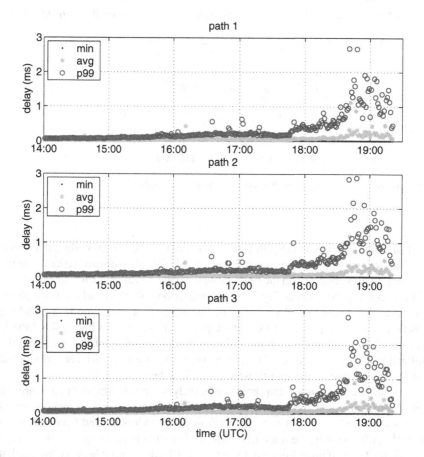

**Fig. 4.11** Minimum, average, and 99th percentile of variable delay per 1-minute interval of Data Set 3.

To investigate how the variable delay evolves throughout the entire trace, we compute the minimum, average, and 99th percentile of variable delay over every 1-minute interval, and the corresponding time series plot is shown in Figure 4.11. The average and 99th percentile delays increase significantly near the last hour of the trace. In case of 99th percentile delays, they often are above 1 ms, which seem to indicate a significant level of congestion during that time period. All three paths experience a heightened level of congestion during almost the same time period that, in turn, makes us suspect the common link link B to be the bottleneck.

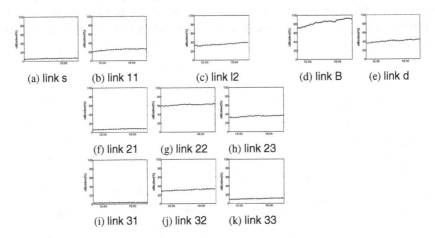

(a) link s        (b) link 11        (c) link l2        (d) link B        (e) link d

(f) link 21        (g) link 22        (h) link 23

(i) link 31        (j) link 32        (k) link 33

**Fig. 4.12** SNMP statistics of link utilization in 5-minute intervals.

The link utilization on the path from the source to the destination should tell us if any link was highly loaded or severely congested. For this purpose, we examine the SNMP statistics collected on the links along the paths to validate what we observe in Figure 4.11. Figure 4.12 displays the link utilization on all links of the three paths. For ease of viewing, we place a plot of link utilization over time in a matching position to its corresponding link in the topology map shown in Figure 4.6. Link B had the highest link utilization of 70% even at the beginning of the trace, and the utilization level increased to 90% near the end of the trace. Clearly, link B was the bottleneck that caused the significant increase in delay.

Another way to confirm that link B was truly the bottleneck is to compare the delay before and after the bottleneck point. We have an extra set of measurements from link12, and can calculate the delay from link s to link 12 and from link 12 to link d[4]. Unfortunately, the trace from link 12 is only 5 hours long, which does not include the last half hour. Figure 4.13(a) has the CCDF of variable delay from link s to link 12. All hourly plots are overlaid on top of each other, meaning that the network condition on the partial path had hardly changed and the packets experienced almost no variable delay (less than 30 $\mu$s for 99.99% of the packets). Figure 4.13(b) shows hourly distributions of variable delay from link 12 to link d. They closely match the shape of Figure 4.9(c). We conclude that the high utilization on link B is the deciding factor that increased variable delay on the path from link s to link d.

---

[4] We do not have measurements from any other intermediate links on Paths 2 or 3.

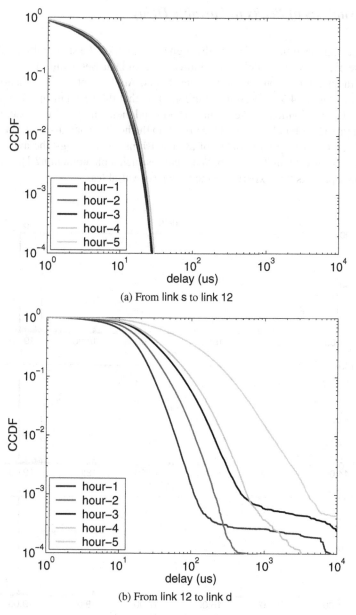

(a) From link s to link 12

(b) From link 12 to link d

**Fig. 4.13** Empirical CCDF of $\{d_{px}^-\}$.

### 4.7.2 Analysis of Peaks in Variable Delay

In the previous section, we identify the high utilization on the shared link, link B, as the cause of large variable delay on all three paths of Data Set 3: thus, the shift in the bulk of the delay distributions in the later hours. Another interesting phenomenon apparent in Figure 4.9(c) (also in Figure 4.10) is that the very tail in several of the hourly delay distributions flattens out abruptly and then tapers off. For example, the very tail (at the probability around $0.5x10^{-3}$) in the hour 1 delay distribution flattens out and reaches a very large delay of almost 10 ms, even though the average link utilization is only about 70% or so. What causes such a phenomenon? This leads us to consider packets that experience the very large variables.

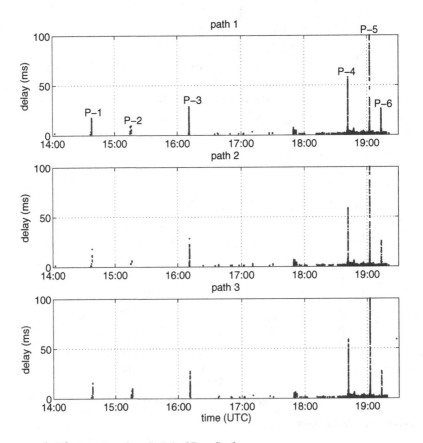

**Fig. 4.14** $\{d^{var}\}$ above 1 ms from Path 1 of Data Set 3.

To examine when and how such a very large delay occurs, we set the threshold of the very large delay at 1 ms, and consider only packets with a variable delay larger than 1 ms. Figure 4.14 is a time series showing the variable delay of these packets vs. the time they arrive at the destination link d. We see that most of these packets appear clustered, and there are six conspicuous peaks, labeled P-1 to P-6, respectively. The maximum delay in each peak exceeds 8 ms and in the case of P-5, reaches up to 100 ms. All the peaks last for a very short period of time, considering the duration of 5.5 hours of our measurement. What causes these peaks? One possible explanation is to attribute them to some random or uncommon events caused, e.g., by router idiosyncrasies [20]. For instance, spiky behavior in an end-to-end packet delay has been previous reported by Ramjee et al. [24]. In their measurements, a delay spike has no ascending slope, but instantly reaches the maximum value and then gradually decreases over time. Claffy et al. observe similar behavior in round-trip time delay and attribute it to Internet hardware problems associated with prefix caching on routers [7]. Before we dismiss these peaks as aberrant behavior, we zoom in on a finer time scale to examine the details of these peaks.

In Figure 4.15, the zoomed-in pictures of the peaks (in the time scale seconds) reveal an interesting behavior. To contrast out packet delays at the peaks, we show packet delays after those peaks for 7 seconds, in Figure 4.16.

Instead of an abrupt rise at the beginning of a peak, the delay gradually increases, reaches a peak, and then decreases. It resembles more of the behavior of a queue building up gradually and eventually being drained. In P-2, it is hard to see the evolution of a peak. P-2 is the shortest, and the dataset may not contain sufficient packets during the P-2 peak formation period to sample the queue building up and draining process. In other peaks, the ascending slope is in general steeper than the descending one.

The most striking observation is from P-5. The delay reaches a plateau at 100 ms and remains at 100 ms for a little longer than 4 s. Delay of 100 ms on an OC-48 link translates to 30 MB of buffer. We do not know the exact buffer size kept at the output queue for link B. Nor do we have packet-level measurements from link B or of other incoming traffic to link B, other than from link 12. At this point, we do not have sufficient information to fully understand the underlying cause of P-5. We can only speculate that the output buffer reached its full capacity and thus the variable delay did not increase any larger.

To see if these peaks occurred on the path segment before the bottleneck, link B, we zoom into the same time period as in Figure 4.15 at link 12. We see no variable delay above 1 ms between link s and link 12. When examining the same time period on the path segment between link 12 and link d, we see peaks very similar to those in Figure 4.14. Figure 4.17 shows the zoomed-in picture corresponding to Figure 4.15. Note that the figure contains only four time series plots; this is because the trace collected on link 12 is shorter than the traces collected on link s and link d, as noted earlier. It is only 5 hours long, and does not cover P-5 or P-6. The four plots in Figure 4.17 match the first four in Figure 4.15 in both height and duration. This suggests that the delay peaks are caused by sudden traffic bursts on the bottleneck link B.

We summarize our findings in this section as follows. The variable delay distribution has two parts, with the first part representing the bulk of the distribution (99th

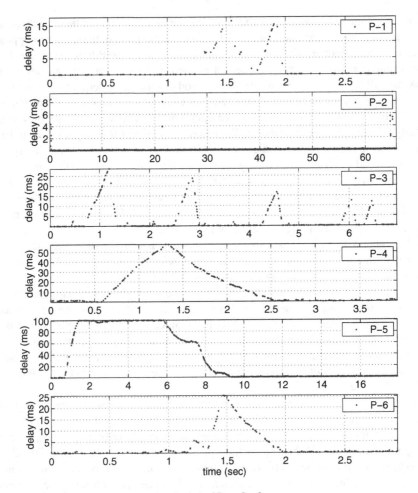

**Fig. 4.15** $\{d^{var}\}$ of peaks, P-1 to P-6 on Path 1 of Data Set 3.

percentile). When the link utilization on a bottleneck link is below 90%, the 99th percentile of the hourly delay distributions remains below 1 ms. Once the bottleneck link is utilized above 90%, the variable delay shows a significant increase overall, and the 99th percentile reaches a few milliseconds. The second part is about the very tail of the distribution. Even when the link utilization is relatively low (below 90%), sometimes a small number of packets may experience delay of an order of magnitude larger than the 99th percentile and affect the shape of the tail, as witnessed by P-1 and P-3 of Figure 4.14. Such a very large delay occurs in clustered peaks, caused by short-term traffic bursts that last only a few seconds and are not captured by the 5-minute average link utilization as reported by SNMP data.

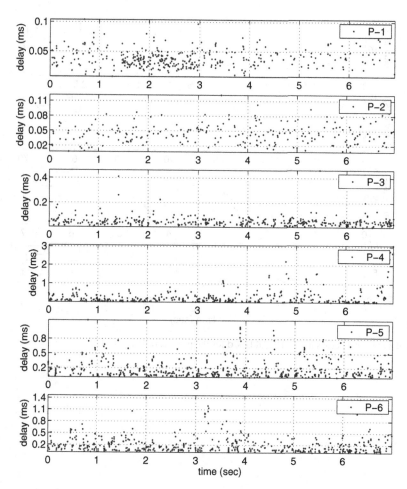

**Fig. 4.16** $\{d^{var}\}$ of 7 seconds after peaks (P-1 to P-6) on Path 1 of Data Set 3.

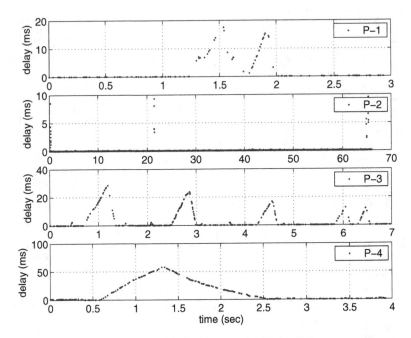

**Fig. 4.17** $\{d^{var}\}$ between link 12 and link d for the periods of P-1 to P-4.

## 4.8 SNMP Data Exploration

In the previous section we showed that the peaks in the variable delay correlate
with utilization levels greater than 90% on the bottleneck link B. This link corresponds to a long-haul link that connects two PoPs inside the Sprint IP backbone
network. Given that such high utilization levels are highly unusual in an operational
IP backbone network, in this section we look into the reasons leading to such a phenomenon. To address this topic we use SNMP data in conjunction with intra- and
inter-domain information.

In Figure 4.18 we present the throughput measurements collected every 5 minutes for link B throughout the month of November. We notice that link B experiences
a jump in its utilization on November 15th, 2002. Traffic on link B increased by 600
Mbps on that particular date and remained at this level until November 21st, 2002
(when our packet trace collection took place). Subsequent days show a significant
drop in traffic that stabilizes around 600 Mbps. In what follows, we explore the
reasons behind the abrupt changes in throughput observed for link B.

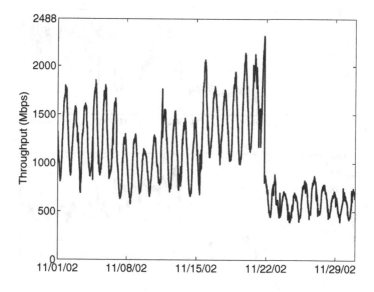

**Fig. 4.18** Throughput measurements for link B throughout the month of November.

Increase in traffic on link B is caused by the multiplexing of the traffic incoming
to the router where link B belongs. Looking into the configuration information for
that particular router, we find that it serves 26 links. Increase in the utilization of
the output link B could be due to two reasons: (i) increase in the overall incoming
traffic, or (ii) a jump in the utilization of one of the input links. Given that the
increase observed occurs as a jump, the second option may be more likely and lead

to the incoming link causing the surge in traffic. To address this issue, we look
into the utilization of all the input links for the same period of time and succeed in
identifying a single link that experiences a jump of equivalent activity.

Backtracking to the router where this link is attached to, we can repeat the same
analysis and attempt to identify one of the input links to this router that manifest the
same 600 Mbps jump on November 15th, 2002. It turns out that in performing this
operation we always manage to identify one single input link manifesting the same
behavior. In that way, we arrive at the access router responsible for the identified
surge of traffic. This router is located on the west coast of the United States and
connects two customers, at OC-12 and Gigabit Ethernet speeds. The first customer
is a large Tier-2 provider, and the other one is a large Internet provider in Asia. The
SNMP repository used for this analysis does not include information for customer
links. However, looking at the 2 OC-192 links connecting this router to the backbone
routers inside the same PoP, we can notice the increase in the total outgoing traffic.
In Figure 4.19 we present the utilization of the two output links of that particular
access router.

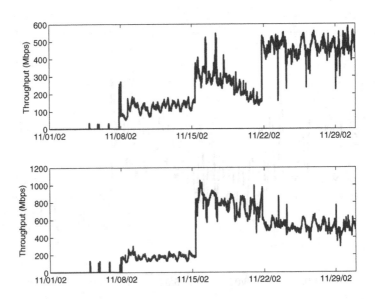

**Fig. 4.19** Throughput measurements for outgoing link of edge router.

Indeed, similar jumps in the utilization of both links can be observed for that
same date. This increase could be due to one or both customers, but a lack of SNMP
information for the customers' links does not allow for such a task. Nevertheless,
we can correlate the time when we witness the abrupt jump in the output utilization
(5 minutes after midnight) to announcements made in the inter-domain routing. We
have installed route listeners inside the Sprint network and record every message

exchanged in the intra- and inter-domain routing. Looking through the BGP logs, we did find that the 2 customers attached to this edge router did advertise 137 and 714 new network prefixes on that location at this same time. New accessible prefixes through this router could lead to additional traffic transiting that access router. Consequently, the increase we witness in the core of the network on link B is due to changes in the inter-domain routing that took place around midnight on November 15th, 2002. We also note that link B was the only link that was negatively impacted by this increase in traffic; all other links in the path had adequate capacity to absorb the surge in traffic without reaching prohibitive levels of utilization.

The inter-domain routing change took place 6 days before we measured the large variable delays through the network. Only 6 days later did the bottleneck link reach 90% utilization. In addition, it maintained this high level of utilization only for 30 minutes. We do see that at approximately 8 pm on November 21st, 2002, there is a sudden drop in the utilization of the bottleneck link and its throughput reduces from 2.3 Gbps down to 800 Mbps.

Link B is an inter-PoP link connecting two PoPs on the east coast of the United States. These two PoPs were interconnected through 4 long-haul links. Before November 21st, 2002, only 2 of those links carried significant amounts of traffic. After the time period, when we observed congestion, we see that the distribution of traffic across the 4 links has changed. Three of the links carry approximately the same amount of traffic, while the fourth link ($L2$) now carries approximately 200 Mbps. These results are presented in Figure 4.20. Such a change in behavior could be due to changes in the ISIS weights for links $L2$ and $L3$. Using the logs of our ISIS router listener for that period of time, we did confirm that such a change in the traffic observed between the two PoPs was due to changes in the intra-domain routing configuration.

In summary, using SNMP and intra- and inter-domain routing information, we showed that: (i) the increase in utilization at the bottleneck link is due to changes in the inter-domain routing and caused a surge of traffic at an access router on the west coast of the United States, (ii) high levels of utilization only persisted for 30 minutes. Network operations resorted to changes in the ISIS configuration for the better load balancing of the traffic between the two PoPs that link B interconnected. Consequently, performance degradation occurred for a small period time and was averted immediately.

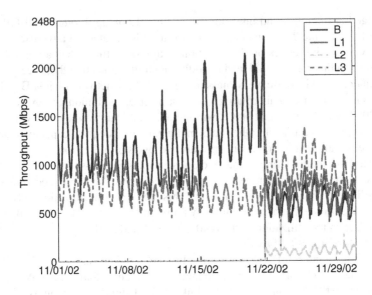

**Fig. 4.20** Throughput of the PoP interconnecting links.

## 4.9 Summary

In this work we present a step-by-step analysis of point-to-point delay from an operational tier-1 backbone network. We first isolate the fixed components in delay, namely, propagation delay, transmission delay, and per-packet overhead at the router. When there are more than one path between two points, we determine the path of each packet, using the minimum delay for each 2-tuple flow and the TTL delta. Once we identify a set of packets that follow the same path, we obtain the minimum path transit time per packet size and subtract it from the point-to-point delay to obtain the variable delay. When the link utilization on all links of the path is under 90%, the 99th percentile of the variable delay remains under 1 ms over 4 to 5 hops and is thus insignificant. However, when a link on the path is utilized over 90%, it becomes a bottleneck in the sense that the weight of the variable delay distribution shifts and even the 90th percentile of the variable delay reaches above 500 $\mu$s and the 99th percentile goes well beyond 1 ms.

Though rare and few in numbers, there are peaks in variable delay often reaching tens of milliseconds in magnitude. We observe such peaks even at below 90% of bottleneck link utilization. We show that these peaks do affect the tail shape of the distribution (above the 99.9th percentile point). We have also investigated the reasons that caused a high traffic load in the bottleneck link. With extensive exploration of SNMP data and intra- and inter-domain routing information, we have found that it is due to changes in the inter-domain routing caused a surge of traffic at an access router.

As we see in this work, many factors contribute to the point-to-point delay in the network. In our observation, point-to-point delays differ as much as 6 ms due to ECMP. Depending on the problem at hand, network engineers can improve current practices by changing a relevant factor.

We believe our observations will be beneficial for network and protocol design in the future. Particularly, our work sheds light on designing effective delay monitoring schemes [5]. For instance, when using active probes for monitoring purposes, network operators and managers should make sure that those probes cover all ECMPs, or at least the longest path, so that they stay informed about the worst-case scenario of their network performance.

# References

1. Abilene Internet2. http://abilene.internet2.edu.
2. bing. Bandwidth pING. http://www.cnam.fr/reseau/bing.html.
3. J.-C. Bolot. End-to-end packet delay and loss behavior in the Internet. In *Proceedings of ACM SIGCOMM*, San Francisco, August 1993.
4. CAIDA. The Cooperative Association for Internet Data Analysis. http://www.caida.org.
5. B.-Y. Choi, S. Moon, R. Cruz, Z.-L. Zhang, and C. Diot. Practical Delay Measurement Methodology in ISPs. In *Proceedings of ACM CoNext (Nominated for best paper award)*, Toulouse, France, Oct. 2005.
6. B.-Y. Choi, S. Moon, Z.-L. Zhang, K. Papagiannaki, and C. Diot. Analysis of Point-To-Point Packet Delay In an Operational Network. *Elsevier Journal of Computer Networks*, 51(13):3812–3827, Sep. 2007.
7. K. C. Claffy, T. Monk, and D. McRobb. Internet tomography. *Nature*, 1999.
8. C. Dovrolis, P. Ramanathan, and D. Moore. What do packet dispersion techniques measure? In *Proceedings of INFOCOM*, Anchorage, AK, April 2001.
9. A. Downey. Using pathchar to estimate internet link characteristics. In *Proceedings of ACM SIGCOMM*, pages 241–250, Cambridge, MA, USA, October 1999.
10. C. Fraleigh, S. Moon, B. Lyles, C. Cotton, M. Khan, D. Moll, R. Rockell, T. Seely, and C. Diot. Packet-level traffic measurements from the Sprint IP backbone. *To be published in IEEE Network*, 2003.
11. U. Hengartner, S. Moon, R. Mortier, and C. Diot. Detection and analysis of routing loops in packet traces. In *ACM SIGCOMM Internet Measurement Workshop*, Marseille, France, November 2002.
12. G. Huston. *ISP Survival Guide: Strategies for Running a Competitive ISP*. John Wiley & Sons, October 1998.
13. Internet End-to-end Performance Monitoring. http://www-iepm.slac.stanford.edu.
14. iperf. Nlanr. http://dast.nlanr.net/Projects/Iperf.
15. V. Jacobson. pathchar. http://www.caida.org/tools/utilities/others/pathchar.
16. M. Jain and C. Dovrolis. End-to-end available bandwidth: Measurement methodology, dynamics, and relation with tcp throughput. In *Proceedings of ACM SIGCOMM*, Pittsburgh, August 2002.
17. S. Kalidindi and M. Zakauskas. Surveyor: An infrastructure for internet performance measurements. In *Internet Networking (INET)*, San Jose, June 1999.
18. Mantra (monitor and analysis of traffic in multicast routers). http://www.caida.org/tools/measurement/mantra.

19. The Measurement and Analysis on the Wide Internet (mawi) working group. http://www. wide.ad.jp/wg/mawi.

20. K. Papagiannaki, S. Moon, C. Fraleigh, P. Thiran, and C. Diot. Measurement and analysis of single-hop delay on an IP backbone network. In *Proceedings of INFOCOM*, San Francisco, CA, April 2002.

21. V. Paxson. End-to-end routing behavior in the Internet. *IEEE/ACM Transactions on Networking*, 5(5):610–615, November 1997.

22. V. Paxson. *Measurement and Analysis of End-to-End Internet Dynamics*, chapter 16. University of California, Berkeley, Ph.D. Thesis edition, April 1997.

23. V. Paxson, A.K. Adams, and M. Mathis. Experiences with nimi. In *Proceedings of Passive and Active Measurement Workshop*, Hamilton, New Zealand, April 2000.

24. R. Ramjee, J. Kurose, D. Towsley, and H. Schulzrinne. Adaptive playout mechanisms for packetized audio applications in wide-area networks. In *Proceedings of INFOCOM*, Montreal, Canada, 1994.

25. RIPE, (Reseaux IP Europeens). http://www.ripe.net/ripe.

26. University of Oregon Route Views Project. http://www.antc.uoregon.edu/route-views.

27. S. Savage. Sting: a tcp-based network measurement tool, proceedings of the 1999 usenix symposium on internet technologies and systems. In *Proceedings of the 1999 USENIX Symposium on Internet Technologies and Systems*, pages 71–79, Boulder, CO, USA, October 1999.

28. W. Stallings. *SNMP, SNMPv2, SNMPv3, and RMON 1 and 2*. Addison Wesley, 3rd edition, 1999.

29. D. Thaler and C. Hopps. Multipath issues in unicast and multicast next-hop selection. Internet Engineering Task Force Request for Comments: 2991, November 2000.

30. TRIUMF Network Monitoring. http://sitka.triumf.ca/.

31. WAND (Waikato Applied Network Dynamics) WITS (Waikato Internet Traffic Storage) Project Network Monitoring. http://wand.cs.waikato.ac.nz/wand/wits/.

# Chapter 5
# Quantile Sampling for Practical Delay Monitoring in Internet Backbone Networks

**Abstract** In this chapter we find a meaningful representation of network delay performance in high quantiles, and use a random sampling technique to estimate them accurately. Our contribution is twofold. First, we analyze router-to-router delay using a large number of traces from the Sprint backbone, and identify high quantiles as a meaningful metric in summarizing a delay distribution. Second, we propose an active sampling scheme to estimate high quantiles of a delay distribution with bounded error. We validate our technique with real data and show that only a small number of probes is needed. We finally show that active probing is the most scalable approach to delay measurement.

**Key words:** delay sampling, passive sampling, active probing, delay measurement, high quantile, delay metrics

## 5.1 Introduction

Point-to-point delay is a powerful "network health" indicator in a backbone network. It captures service degradation due to congestion, link failure, and routing anomalies. Obtaining meaningful and accurate delay information is necessary for both ISPs and their customers. Thus delay has been used as a key parameter in Service Level Agreements (SLAs) between an ISP and its customers [5, 18]. In this chapter, we systematically study how to measure and report delay in a concise and meaningful way for an ISP, and how to monitor it efficiently.

Operational experience suggests that the delay metric should report the delay experienced by most packets in the network, capture anomalous changes, and not be sensitive to statistical outliers such as packets with options and transient routing loops [1, 4]. The common practice in operational backbone networks is to use ping-like tools. ping measures network round trip times (RTTs) by sending ICMP requests to a target machine over a short period of time. However, ping was not designed as a delay measurement tool, but a reachability tool. Its reported delay includes uncertainties due to path asymmetry and ICMP packet generation times at routers. Furthermore, it is not clear how to set the parameters of measurement tools (e.g., the test packet interval and frequency) in order to get a certain accuracy.

Inaccurate measurement defeats the purpose of performance monitoring. In addition, injecting a significant number of test packets for measurement may affect the performance of regular traffic, as well as tax the measurement systems with unnecessary processing burdens. More fundamentally, defining a metric that can give a meaningful and accurate summary of point-to-point delay performance has not been considered carefully.

We raise the following practical concerns in monitoring delays in a backbone network. How often should delay statistics be measured? What metric(s) capture the network delay performance in a meaningful manner? How do we implement these metrics with limited impact on network performance? In essence, we want to design a practical delay monitoring tool that is amenable to implementation and deployment in high-speed routers in a large network, and that reports useful information.

The major contributions of this chapter are three-fold: (i) By analyzing the delay measurement data from an operational network (Sprint US backbone network), we identify high-quantiles [0.95-0.99] as the most meaningful delay metrics that best reflect the delay experienced by most of packets in an operational network, and suggest 10-30 minute time scale as an appropriate interval for estimating the high-quantile delay metrics. The high-quantile delay metrics estimated over such a time interval provide a best representative picture of the network delay performance that captures the major changes and trends, while they are less sensitive to transient events, and outliers. (ii) We propose and develop an active probing method for estimating high-quantile delay metrics. The novel feature of our proposed method is that it uses the minimum number of samples needed to bound the error of quantile estimation within a prescribed accuracy, thereby reducing the measurement overheads of active probing. (iii) We compare the network wide overhead of active probing and passive sampling for delays. To the best of our knowledge, this is the first effort to propose a complete methodology to measure delay in operational networks and validate the performance of the active monitoring scheme on operational data.

The remainder of this chapter is organized as follows. In Section 5.2 we provide the background and data used in our study. In Section 5.3 we investigate the characteristics of point-to-point delay distributions obtained from the packet traces and discuss metrics used in monitoring delay in a tier-1 network. In Section 5.4 we analyze how sampling errors can be bounded within pre-specified accuracy parameters in high quantile estimation. The proposed delay measurement scheme is presented and its performance is evaluated using packet traces in Section 5.5. We compare the overheads of our active probing and a passive sampling in Section 5.6. We summarize the chapter in Section 5.7.

## 5.2 ISP Delay Measurement: Background and Methodology

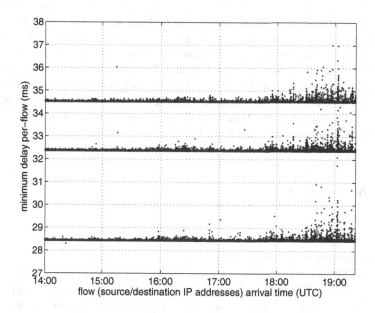

**Fig. 5.1** Presence of ECMP in Data Set 3.

We describe our data set and provide some background about point-to-point delay observed from this data.

### 5.2.1 Data

We have collected packet traces from Sprint's tier-1 backbone using the methodology described in [3]. The monitoring system passively taps the fibers to capture the first 44 bytes of all IP packets. Each packet header is timestamped. The packet traces are collected, from *multiple* measurement points *simultaneously*, and span over a *long period* of time (e.g. hours). All the monitoring systems are synchronized by GPS (Global Positioning System). The resolution of the clock is sub-microsecond, allowing us to disambiguate packet arrival times on OC-48 links. The timestamp maximum error is 5 microseconds.

To obtain packet delays between two points, we first identify packets that traverse two points of measurements. We call this operation *packet matching*. We use hashing to efficiently match two packet traces. We use 30 bytes out of the first 44 bytes in the hash function. The other 14 bytes are IP header fields that would not help disambiguate similar packets (e.g. version, TTL, and ToS). We occasionally find duplicate packets. Since these packets are totally identical, they are a source of error in the matching process. Given that we observe less than 0.05% of duplicate packets in all traces, we remove these duplicate packets from our traces.

We have matched more than 100 packets traces, and kept only those *matched trace* that exhibited many (more than half a million) successful matched packets. The matched traces are from paths with various capacities and loads over multihop nodes. For a succinct presentation, we have chosen to illustrate our observations of with 3 matched traces out of the 21 we studied. The traces shown are representative and the other traces show similar results. The statistics of these three matched trace are shown in Table 5.1. In all the matched trace data sets, the source and destination links are located on the West Coast and the East Coast of the United States respectively, rendering trans-continental delays over multiple hops.

## 5.2.2 Background

**Table 5.1** Summary of matched traces (delay in ms)

| Set | From | To | Period | Packets | min. | Avg. | med. | .99$^{th}$ | max. |
|---|---|---|---|---|---|---|---|---|---|
| 1 | OC-48 | OC-12 | 16h 24m | 1,349,187 | 28.430 | 28.460 | 28.450 | 28.490 | 85.230 |
| 2 | OC-12 | OC-12 | 5h 27m | 882,768 | 27.945 | 29.610 | 28.065 | 36.200 | 128.530 |
| 3 | OC-12 | OC-48 | 5h 21m | 3,649,049 | 28.425 | 31.805 | 32.425 | 34.895 | 135.085 |

We now briefly discuss the characteristics of actual packet delays observed on the Sprint US IP backbone that was detailed in Chapter 4.

The empirical cumulative probability distributions of point-to-point delays using a bin size of 5 $\mu$s is shown Figure 5.2. For ease of observation, we divide the duration of traces into 30 minute intervals and then plot distributions for the first and last three intervals of each trace duration.

Delay distributions exhibit different shapes, as well as change over time, especially in Data Set #2 and #3. We explain these differences as follows. In theory, the packet delay consists of three components: propagation delay, transmission delay and queueing delay. Propagation delay is determined by the physical characteristics of the path. Transmission delay is a function of the link capacities along the path as well as the packet size. Queueing delay depends on the traffic load along the path, and thus varies over time. In practice, other factors add variations to the delay packets experience in an operational network. First, Internet packet sizes are known to exhibit three modes, where the peaks are around 40, 576 (or 570), and 1500 bytes [7]. When there is little queueing on the path, the packet size may impact the shape of a distribution even in the multi-hop delays, as shown in Figure 5.2(a). In addition, routing can introduce delay variability. Route may change over time because of link failure. Figure 5.2(b) shows that the path between the two measurement points changed within the last 30 minutes. Furthermore, packets can take multiple routes between two points because of load balancing, as in Figure 5.2(c). Equal-cost multi-path (ECMP) routing [19] is commonly employed in operational networks. Routers (e.g., Cisco routers in our study) randomly split traffic using a hash function that takes the source and the destination IP addresses, and the router ID (for traffic splitting decision to be independent from upstream routers) as input to determine the outgoing link for each packet. Therefore packets with the same source and destination IP addresses always follow the same path. We define a *(two-tuple) flow* to be a set of packets with the same source and destination IP addresses, and group packets into flows. We then compute the *minimum* packet delay for each flow. As suggested in Chapter 4, if the two flows differ significantly in their minimum delays, they are likely to follow two different paths. In Figure 5.1 we plot the minimum delay of each flow by the arrival time of the first packet in the flow for Data Set 3. The plot demonstrates the presence of three different paths, each corresponding to one step in the cumulative delay distribution of Figure 5.2(c). Last, extreme packet delays may occur even under a perfectly engineered network, due to routing loops [4] or router architecture [1] related issues. From the perspective of a practical delay monitoring, we need to take all these factors into account to provide an accurate and meaningful picture of actual network delay.

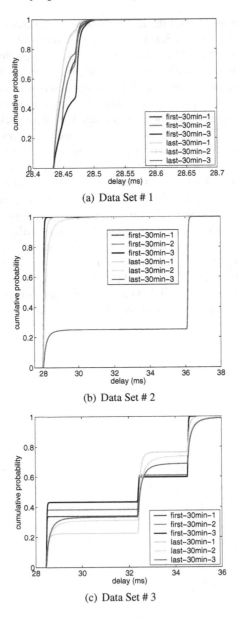

(a) Data Set # 1

(b) Data Set # 2

(c) Data Set # 3

**Fig. 5.2** Empirical cumulative probability density function of delay over 30 minute interval

## 5.3 Measured Delay Distributions and Their Implications

Using the precise point-to-point packet delay measurements obtained from packet traces, in this section we analyze the characteristics of measured delay distributions and discuss their implications in the design of effective delay measurement tools using active probing (i.e., sampling). In particular, we investigate how various network factors contribute to the point-to-point packet delays and make delay distributions in an operational network difficult to characterize using standard statistical metrics such as mean and variance. Our observation raises a more fundamental question that is of both theoretical and practical importance in network delay measurement: what are the *meaningful* delay metrics that can provide a representative picture of the network delay performance, and thus are useful to an ISP for network engineering purposes? In section 5.3.2 we propose and argue that *high quantile* is such a meaningful delay metric.

### 5.3.1 Analysis of Measured Delay Distributions

From the point-to-point packet delay measurements from the three trace pairs listed in Table 5.1, we obtain the empirical probability density function (pdf) of point-to-point delays, using a bin size of 1 $\mu$s. The resulting empirical probability density functions for the three trace pairs are shown in Figure 5.3. First of all, from the figures we see that the measured delay distributions look drastically different from each other, and none of them can be well characterized by standard statistical distributions such as normal distributions. Delay distributions from other trace pairs we have analyzed show similar properties. In the following, we examine each delay distribution more closely and investigate the plausible network factors that may have contributed to the vastly different shapes of these delay distributions.

First, we observe that the delay distribution in Figure 5.3(a) has three "peaks" and they are only a few tens of microseconds apart. Because of the closeness of the three peaks, we conjecture that the peaks are likely caused by the typical IP packet size distribution that is known to exhibit a tri-modal distribution [14]. To validate this conjecture, we classify packets into three classes based on their size: smaller than 400 bytes, larger than 800 bytes, and those in between. Figure 5.4 plots the corresponding empirical probability density function of the point-to-point delays for packets in each class. The empirical pdf of each class now shows a uni-modal distribution, the peak of which corresponds to one of the three peaks in Figure 5.3(a). Hence, we conclude that the tri-modal delay distribution in Figure 5.3(a) is likely due to the effect of tri-modal packet size distribution. It is interesting to note that it is shown in [14] that the single-hop delay distributions typically exhibit tri-modal distributions, due to the same effect of tri-modal packet size distribution. That the packet sizes have a dominant effect on single-hop delay distribution is attributed to the generally very low link utilization in a well-provisioned tier-1 carrier network, where relatively long queueing delay rarely occurs in high-speed routers. Figure 5.3(a) shows that even across multiple hops, packet sizes can play an important role when there is very little queueing along the path, and transmission delay is the dominant variable in point-to-point packet delays. In fact, many other trace

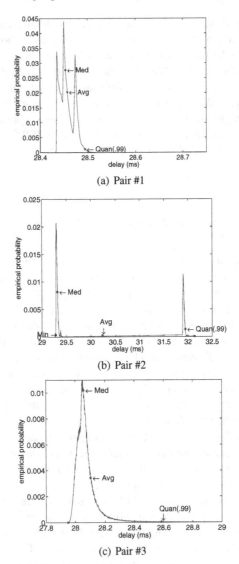

**Fig. 5.3** Empirical probability density functinos of delay and the related statistics.

pairs we studied also exhibited the tri-modal delay distribution, and it is one of most common distributions[1].

Compared with Figure 5.3(a), the delay distribution in Figure 5.3(b) has a very different form: it exhibits a bi-modal distribution with one peak at 29.3 ms and the other at 31.9 ms; in addition, the two peaks are significantly apart, with a distance

---

[1] See http://ipmon.sprint.com for the longer list of empirical pdfs of delay.

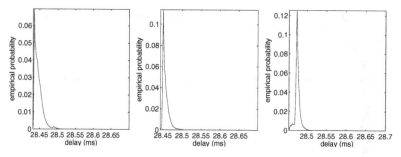

**Fig. 5.4** Empirical probability density function of delay for three different packet size classes.

greater than 2 ms. This rules out the possible effect of packet sizes, and points to another network factor as the plausible culprit – that the packets in this trace pair may have taken two different physical paths between the ingress and the egress points, one with a longer propagation delay than the other. To investigate, we first check whether there is a route change in the duration of the trace collection by plotting the minimum, mean, and maximum delay per minute over time. If there had been a route change, we should see an abrupt change of 2.4 ms in the minimum delay at some point. However, we do not observe such a change in the time-series plot. This leads us to speculate that equal cost multi-path (ECMP) routing must be used between the ingress and egress points. In the carrier network where the packet traces are collected, ECMP routing is often used for load balancing and providing high availability in the face of a network element failure. To maintain in-order delivery within a flow (e.g., a TCP flow), traffic is split onto ECMPs based on the source IP address, the destination IP address, and the router identity. Thus, packets that have the same source and destination IP addresses follow the same path, and are delivered in order. To validate this conjecture, we group packets into "five-tuple" flows based on their source IP address, source port, destination IP address, destination port, and protocol number. The mean delay for each flow is plotted in Figure 5.5. The figure reveals that per-flow mean delays centered around two separate lines, one around 29.3 ms and another around 31.9 ms. They correspond to the two peaks in Figure 5.3(b). This confirms our conjecture that the bimodality in the delay distribution is caused by ECMP routing: two physical paths are used in routing packets between the ingress and egress points. In general, we expect that if there are more than two ECMPs between two points of measurement, then the corresponding number of peaks (or "modes") will appear in the resulting point-to-point delay distribution.

We now turn our attention to the delay distribution in Figure 5.3(c), which has a uni-modal distribution. The absence of "widely" separated modes implies that a single path is likely used between the ingress and egress points. Furthermore, the bulk of the distribution covers a range between 27.95 ms to 28.25 ms, with a span of about 0.3ms. Compared with Figure 5.3(a), this seems to indicate that whatever effect the packet size distribution may have on the delay distribution. It may have been masked by the bulk of the delay distribution. The large bulk and relatively long tail of the distribution suggests queueing delay as a likely factor that

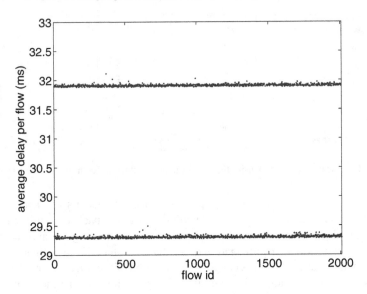

**Fig. 5.5** Per-flow minimum delay of trace pair #2.

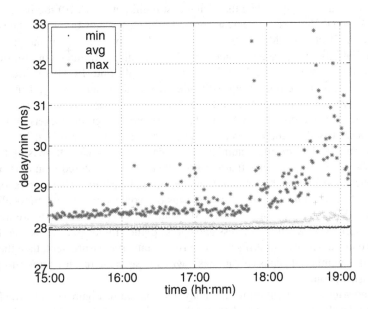

**Fig. 5.6** Time-series plot of the minimum, the mean, and the maximum delay per minute of trace pair #3.

influences the shape of the overall point-to-point delay distribution. To discern the effect of queueing delay caused by traffic load increases, we plot the time-series of the minimum, the mean, and the maximum delays per minute in Figure 5.6. The gradual increase in the mean and the maximum delay per minute in the second half of the graph underscores the impact of queueing in the resulting overall delay distribution.

To summarize, we have shown that point-to-point delay distributions in an operational network can exhibit vastly different forms. The shape of delay distributions can be affected by many different network factors such as packet sizes, traffic loads along the path, and ECMP routing. In general, these delay distributions cannot be characterized by standard statistical distributions such as normal distribution. Clearly it is not straightforward to find a parametric model that characterizes all possible shapes of delay distribution. These observations have significant implications in the design of delay measurement tools using active probing, and more fundamentally, in what delay metrics we should use for a meaningful representation of the network delay performance. In the next subsection we will discuss the latter issue first. Then in Section 5.5 we will address the problem of how to design effective delay measurement tools using high quantile estimation.

### 5.3.2 Metrics Definition for Practical Delay Monitoring

The objective of our study is to design a practical delay monitoring tool to provide a network operator with a *meaningful* and *representative* picture of delay performance of an operational network. Such a meaningful and representative picture should tell the network operator *major* and *persistent* changes in delay performance (e.g., due to persistent increase in traffic loads) *not* transient fluctuations due to minor events (e.g., a transient network congestion). Hence in designing a practical delay monitoring tool, we need to first answer two inter-related questions: (i) what metrics should we select so as to best capture and summarize the delay performance of a network, namely, by a majority of packets; and (ii) over what time interval should such metrics be estimated and reported? We refer to this time interval as the (metrics) *estimation* interval. Such questions have been studied extensively in statistics and performance evaluation (see [8], for a general discussion of metrics in performance evaluation). From the standpoint of delay monitoring in an operational network, we face some unique difficulties and challenges. Thus our contribution in this respect lies in putting forth a practical guideline through detailed analysis of delay measurements obtained from Sprint's operational backbone network: we suggest *high quantiles ([0.95,0.99]) estimated over a 10-30 minute time interval* as meaningful metrics for ISP practical delay monitoring. In the following we present our analysis and reasoning using the three data sets discussed in the previous section as examples.

To analyze what metrics provide a meaningful and representative measure of network delay performance, we consider several standard metrics, i.e., minimum,

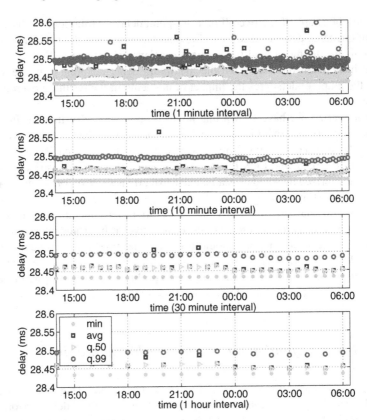

**Fig. 5.7** Delay metrics over different estimation intervals (Data Set #1).

average, maximum, median (50% percentile, or 0.5th quantile) and high quantiles (e.g., 0.95th quantile), estimated over various time intervals (e.g., 30 seconds, 1 minute, 10 minutes, 30 minutes, 1 hour), using the delay measurement data sets collected from the Sprint operational backbone networks. Results are plotted in Figures 5.7~5.9. Note that here we do not plot the maximum delay metrics as maximum delays are frequently so large that they obscure the plots for the other metrics. Some statistics of the maximum delays are given in Table 5.1, where we see that maximum delays can be several multiples of the $0.99^{th}$ quantiles.

From the figures, we see that delay metrics estimated over small time intervals (e.g., 1-minute) tend to fluctuate frequently, and they do not reflect significant and persistent changes in performance or trends (for example, Figure 5.7, Figure 5.8 at time 14:40 and Figure 5.9 at time 16:30).[2]

On the other hand, the increase in delay around 18:30 and onwards in both Data Set #2 and Data Set #3, represents a more significant change in the delay trend,

---

[2] We do not know exactly what caused the delays. We focus our work on *measuring* and *estimating* delays. and investigating reasons of the delay is out of the scope in our work.

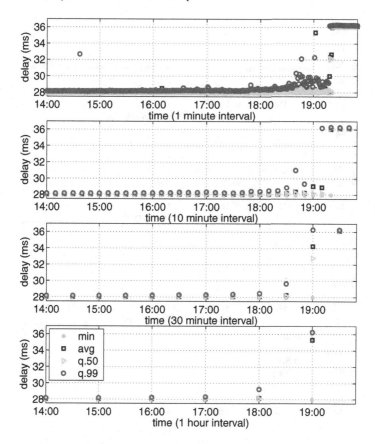

**Fig. 5.8** Delay metrics over different estimation intervals (Data Set #2).

and should be brought to the attention of network operators. Note also that in a few occasions the average delays particularly *estimated over a small time interval* are even much larger than the $0.99^{th}$ quantiles (see, the top two plots in Figure 5.7 around 18:00 and 21:00) – this is due to the extreme values of the maximum delays that drastically impact the average.

As a general rule of thumb, the time interval used to estimate delay metrics should be large enough not to report transient fluctuations, but not too large in order to capture in a timely fashion the major changes and persistent trends in delay performance. In this regard, our analysis of the data sets suggests that 10-30 minute time interval appear to be an appropriate delay estimation interval. As an aside, we remark that our choice of 10-30 minute time interval is also consistent with the studies of others using different measurement data. For example, the active measurement study in [20] using NIMI measurement infrastructure [15] has observed that in general packet delay on the Internet appears to be steady on time scales of 10-30 minutes.

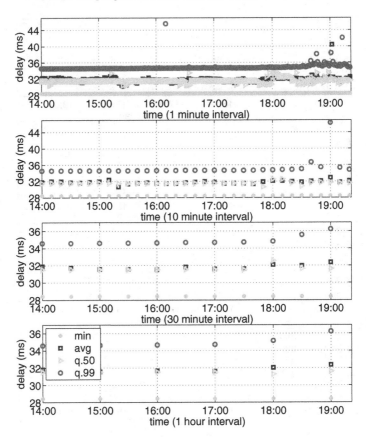

**Fig. 5.9** Delay metrics over different estimation intervals (Data Set #3).

In choosing delay metrics, similar properties are desired. A meaningful metric to ISPs should characterize the delay experienced by most of packets, thereby providing a good measure of the typical network performance experienced by network users. Furthermore, such a delay metric should not be too sensitive to outliers. We summarize the pros and cons of various delay metrics as below:

- Maximum delay suffers greatly from outliers. Some packets might experience extreme delays even under well-managed and well-provisioned networks [4, 6, 10] due to IP packets with options, malformed packets, router anomalies and transient routing loop during a convergence time. The rate of outliers is such that there would be such a packet in almost every time interval. However, packets that experience the maximum delay are not representative of the network performance.
- Average or median delay have the main disadvantage of not capturing delay variations due to route changes (Figure 5.2(b)) or load-balancing (Figure 5.2(c)) that

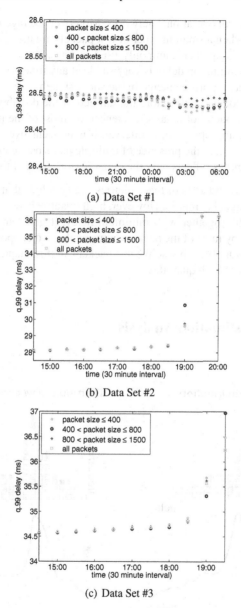

(a) Data Set #1

(b) Data Set #2

(c) Data Set #3

**Fig. 5.10** Impact of packet size on quantile (30 minute estimation interval)

happen frequently in operational networks. Moreover, average is sensitive to outliers especially when a small number of test packets are used.

- Minimum delay is another commonly used metrics. We can see from Figures 5.7~5.9 that the minimum delay is very stable at any time granularity. A change in minimum delay reports a change in the shortest path.
- High quantiles ([0.95, 0.99]) ignore the tail end of the distribution and provides a practical upper bound of delay experienced by most of the packets. When estimated over the appropriate time interval, it is not sensitive to a small number of outliers. However, in the presence of multiple paths between the measurement points, high quantiles reflects only the delay performance of the longest path.

Weighing in the pros and cons of these metrics, we conclude that high percentile is the most meaningful delay metric. However, high quantile does not detect a change in the shortest path. Together with minimum delay, it gives an ISP the range of delays experienced by most of the packets between the two endpoints. As minimum delay is easy to capture [9] using active test packets, in this chapter, we focus on the accurate estimation of high quantiles.

## 5.4 Quantile Estimation Analysis

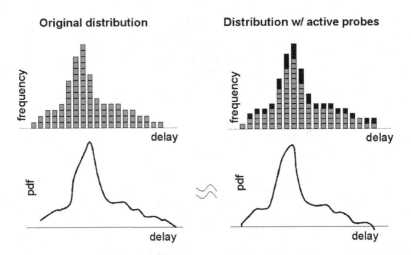

**Fig. 5.11** Distribution with active probes.

In this section we develop an efficient and novel method for estimating high-quantile delay metrics: it estimates the high-quantile delay metrics within a prescribed error bound using a number of required test packets. In other words, it attempts to minimize the overheads of active probing. In the following, we first

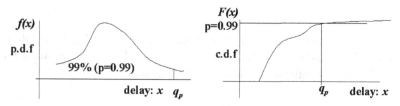

**Fig. 5.12** Percentile of a distribution.

formulate the quantile estimation problem and derive the relationship between the number of samples and the estimation accuracy. Then, we discuss the parameters involved to compute the required number of samples.

We derive the required number of test packets to obtain a pre-specified accuracy in the estimation using Poisson modulated probing. Active test packets perform like passive samples under the following two assumptions. First, the amount of test packets should be negligible compared to the total traffic, so that it does not perturb the performance it measures (See Figure 5.11). Second, the performance of test packets should well represent the performance of regular traffic. Both assumptions are held, which rationalizes our use of active probing. As we will see later, the required number of test packets is relatively small, thus it is negligible on today's high speed backbone networks. Also, we encapsulate the test packets in regular UDP packets so that they do not receive special treatments in a router, unlike packets with IP option or ICMP packets that go to the slow-path of a router.

Now, we formally define a quantile of a delay distribution. Let $X$ be a random variable of delay. We would like to estimate a delay value $q_p$ such that the 99% (*i.e.*, $p = 0.99$) of time, $X$ takes on a value smaller than $q_p$. The value $q_p$ is called the $p^{th}$ quantile of delay and is the value of interest to be estimated. It is formally stated as[3]:

$$q_p = \inf\{q : F(q) \geq p\} \qquad (5.1)$$

where $F(\cdot)$ denotes a cumulative probability density function of delay $X$.

Suppose we take $n$ random samples, $X_1, X_2, \ldots, X_n$. We define $\hat{F}$, an empirical cumulative distribution function of delay, from $n$ samples ($i = 1, \ldots, n$) as

$$\hat{F}(q_p) = \frac{1}{n} \sum_{i=1}^{n} I_{X_i \leq q_p} \qquad (5.2)$$

where the indicator function $I_{X \leq q_p}$ is defined as

$$I_{X_i \leq q_p} = \begin{cases} 1 & \text{if } X_i \leq q_p, \\ 0 & \text{otherwise.} \end{cases} \qquad (5.3)$$

Then, the $p^{th}$ sample quantile is determined by

---

[3] Note that theoretically, the original delay distribution can be considered as a continuous function, and the measured delay distribution is a realization of it.

$$\hat{q}_p = \hat{F}^{-1}(p) \tag{5.4}$$

Since $\hat{F}(x)$ is discrete, $\hat{q}_p$ is defined using order statistics. Let $X_{(i)}$ be the $i$th order statistic of the samples, so that $X_{(1)} \le X_{(2)} \le \cdots \le X_{(n)}$. The natural estimator for $q_p$ is the $p^{th}$ sample quantile ($\hat{q}_p$). Then, $\hat{q}_p$ is computed by

$$\hat{q}_p = X_{(\lceil np \rceil)} \tag{5.5}$$

Our objective is to bound the error of the $p^{th}$ quantile estimate, $\hat{q}_p$. See Figure 5.12 for the illustration. More specifically, we want the absolute error in the estimation $|\hat{q}_p - q_p|$ to be bounded by $\varepsilon$ with high probability of $1 - \eta$:

$$Pr\left\{|\hat{q}_p - q_p| > \varepsilon\right\} \le \eta \tag{5.6}$$

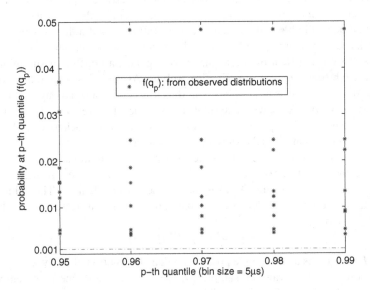

**Fig. 5.13** Empirical tail probability.

Now we discuss how many samples are required to guarantee the pre-specified accuracy using random sampling. Since they are obtained by random sampling, $X_1, X_2, \ldots, X_n$ are i.i.d. (independent and identically distributed) samples of the random variable $X$. It is known that quantile estimates from random samples asymptotically follow a normal distribution as the sample size increases (See Appendix for details).

$$\hat{q}_p \xrightarrow{D} N\left(q_p, \frac{\sigma^2}{n}\right) \text{ where } \sigma = \frac{\sqrt{p(1-p)}}{f(q_p)} \tag{5.7}$$

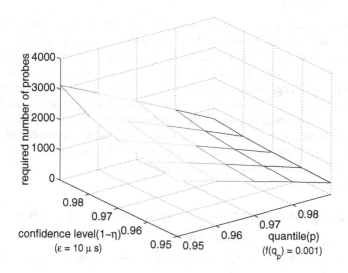

**Fig. 5.14** Number of test packets required ($\varepsilon = 10\mu s$, $f(q_p) = 0.001$).

$f(q_p)$ is the probability density at the $p^{th}$ quantile of the actual distribution. Eq. (5.7) is called Bahadur expression [17]. The estimator is known to have the following properties: (i) *unbiasedness*: the expectation of the estimate is equal to the true value (i.e., $E(\hat{q}_p) = q_p$). (ii) *consistency*: As the number of test packets $n$ increases, the estimate converges to the true value (i.e., $\hat{q}_p \to q_p$ as $n \to \infty$). Note that the above analysis is based on *random* sampling. Thus the analysis of accuracy such as confidence interval ($\varepsilon$) and confidence level ($1 - \eta$) is applicable *regardless* of the underlying delay distribution from the Central Limit Theorem.

We derive from Eq. (5.6) and (5.7) the required number of samples to bound the estimation error within the pre-specified accuracy as

$$n^* = \left\lceil z_p \cdot \frac{p(1-p)}{f^2(q_p)} \right\rceil \tag{5.8}$$

where $z_p$ is a constant defined by the error bound parameters (i.e., $z_p = \left(\frac{\Phi^{-1}(1-\eta/2)}{\varepsilon}\right)^2$), and $\Phi(\cdot)$ is the cumulative probability function of standard normal distribution.

Eq. (5.8) concisely captures the relationship of the number of samples on the quantile of interest ($p$), the accuracy parameters ($\varepsilon, \eta$) and a parameter of original delay distribution ($f(q_p)$).

From Eq. (5.7) and (5.8), we show that the variance of the estimate is bounded as

$$Var(\hat{q}_p) = \frac{p(1-p)}{f^2(q_p) \cdot n^*} \le \frac{1}{z_p} \tag{5.9}$$

since $n^* \geq z_p \cdot \left( \frac{p(1-p)}{f^2(q_p)} \right)$.

The derivation here is for cases with low sample fractions from a large population. We have analyzed the results as if we sampled with replacement though the actual sampling is done without replacement, as it makes the formula simple and enables us to compute the required number of samples concisely. When the sampling fraction is non-negligible, an extra factor should be considered in computing the number of samples. The impact is that the actual variance from the sampling without replacement would be smaller than the one from with replacement. Thus, the actual estimation accuracy achieved is higher with the given number of samples. Practically the analysis of sampling with replacement is used as long as the population is at least 10 times as big as the sample [12].

**Send a probe when a random timer** $t$ **(0~T)  expires**

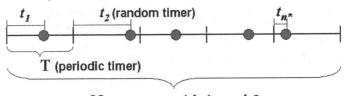

**Fig. 5.15** Scheduling $n^*$ pseudo-random samples.

Unfortunately, $f(q_p)$ is not known in practice. Therefore, it can only be approximated. The required number of samples is inversely proportional to $f^2(q_p)$.

A reasonable lower-bound of the value should be used in the computation of $n^*$, so that the accuracy of the quantile can be guaranteed. We investigate an empirical values of $f(q_p)$ using our data. The empirical p.d.f. of a delay distribution should be evaluated in terms of a time granularity of measurements. As the bin size or the time granularity of distribution gets larger, the relative frequency of delay becomes larger. In order to approximate $f^2(q_p)$, we observe the tail probabilities of delay distributions from the traces. However, for 10-30 minute durations of various matched traces from differing monitoring locations and link speeds , we find that the probabilities at high quantiles, $f(q_p), (0.95 \leq p \leq 0.99)$ vary little and can be reasonably lower bounded. Figure 5.13 shows the probability of high quantiles of the matched traces at time granularity of $5\mu s$. We find the values between 0.0005 to 0.001 are sufficient as the lower-bound of the tail probability for quantiles of $0.95 \leq p \leq 0.99$. Meanwhile, if $p$ approaches to 1 (e.g., $p = 0.99999$), the quantile is close to the maximum and $f(q_p)$ becomes too small requiring large number of samples. Note that when the tail probability becomes heavier, $f(q_p)$ becomes larger making the estimate more accurate. On the other hand, when the tail probability becomes smaller than the approximated, the accuracy of an estimate (the variance of estimation) would not degrade much, since the variance of the original packet delay

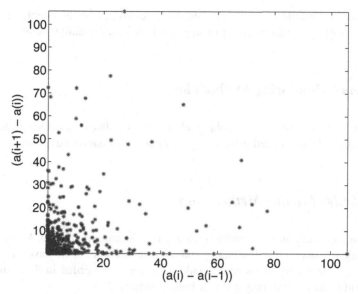

**Fig. 5.16** Correlation of inter-packet time of long delayed packets (correlation coefficient = $1.8e - 6$).

would be small. Therefore, with given accuracy parameters and the lower bound of $f(q_p)$, the number of test packets is decided as a constant.

Figure 5.14 shows the number of required samples for different quantiles and different accuracy parameters[4]. It illustrates the degree of accuracy achieved with the number of samples, and thus provides a guideline on how to choose the probing frequency for a given quantile $p$ to be estimated. A sample size between a few hundred and a few thousand test packets ($420 \sim 3200$) is enough for ($\varepsilon = 10\mu s, 1 - \eta \in [0.95, 0.99]$) range of accuracy and ($q_{.95} \sim q_{.99}$) high quantile. With high speed links (1 Gbps and above), we consider the amount of injected traffic for probing purpose negligible compared to the total traffic. For example, 1800 packets over a 10 minute period corresponds to about 3 packet per second on average. Suppose 64 byte packets are used for the test packets. This would constitute only 1.5 Kbps which is 0.0002% of the total traffic for a 30% loaded OC-48 link.

Before leaving this section we comment on estimating an entire distribution, even though our focus in this chapter is on a point estimation of a most representative delay metric. Note that Eq. (5.8) applies to any quantile in a distribution. Thus, the estimated quantiles enjoy the pre-scribed accuracy, if the minimum required number of samples for the quantile, $n^*$ is smaller than the used number of samples. In particular, as the quantile goes closer to median ($q = 0.5$) and the probability

---

[4] Note that $f(q_p)$ for each high quantile is fixed equally as 0.001 from empirical observation shown in Fig. 5.13.

density at the quantile $f(q_p)$ gets larger, the required number of samples becomes smaller, resulting that the accuracy of an estimation for the quantile becomes higher.

## 5.5 Delay Monitoring Methodology

In this section, we describe our probing scheme and validate its accuracy using delay measurement data collected from the Sprint operational backbone network.

### 5.5.1 Active Probing Methodology

The design goal of our active probing scheme is to estimate high quantile effectively and efficiently over a fixed estimation interval. In Section 5.4 we have shown that at least $n^*$ number of independent random samples are needed in the estimation interval in order to accurately estimate high quantiles.

We proceed as follows. To generate $n^*$ number of test packets within an estimation interval $I$, we divide the interval into $n^*$ subintervals of length $T(= I/n^*)$. With the help of two timers – a periodic $(T)$ timer and a random $(t \in [0,T])$ one, a random test packet is generated for each subinterval $T$ in a time-triggered manner (i.e., whenever a random timer $t$ expires, a test packet is generated). At the end of an estimation interval $(I)$, the delay quantile of the test packets is computed and reported. Figure 5.15 illustrates graphically how to generate the pseudo-random test packets. With this scheme, we ensure that $n^*$ number of test packets are generated independently in every estimation interval without generating a burst at any moment.

We now verify if our time-triggered *pseudo*-random probing performs close to random sampling in estimating high delay quantile. If the inter-arrival times of packets with long delays (e.g., $0.95^{th}$ quantile or larger) are temporarily correlated, the pseudo-random probing would not enable us to estimate high percentile delay well. However, we find that the correlation coefficient is close to 0 (for other intervals and traces with the estimation interval of 10-30 minutes). If the arrival times of packets with long delays (e.g., $.95^{th}$ quantile or larger) are temporarily correlated, the pseudo-random probing may not capture the delay behavior well. Figure 5.16 shows the scatter plot of inter-arrival times of packets with long delays (for the last 30 minutes of Data Set #3). It illustrates that inter-arrival times of packets with long delays are essentially independent.

Test packets scheduling aside, there are several practical issues in implementing a probing scheme such as protocol type and packet size. For the type of test packets, we choose to use UDP packets instead of ICMP packets that are used in `ping`-like active probing softwares. ICMP packets are known to be handled with a lower priority at a router processor. Thus their delay may not be representative of actual packet delay. Test packet size might affect the measure of the delay. We analyzed all matched traces and found that packet size has little impact on high quantile. This is best illustrated in Figure 5.10 where we classify packets into three clusters based on the packet sizes, and computed their $.99^{th}$ quantile, compared with that of all packets. As observed, high quantiles from individual packet size classes are similar, and one particular packet size class does not reflect high quantile from all packets better consistently. It provides the evidence that high quantile delays are not likely to come from packets of a large size, thus the size of test packet should not impact the accuracy of high quantile estimation.

We also have performed a thorough analysis of packet properties in order to detect a correlation between packet fields and delay, if any. However, we did not find any correlation between packet types and the delay distribution. This result confirms that the tail of distribution comes from queueing delay rather than due to a special packet treatment at routers.

As ECMP is commonly employed in ISPs, we need to make sure that our test packets take all available paths when they exist. Load balancing is done on a flow basis, in order to preserve packet sequence in a flow. Therefore, we propose to vary the source address of test packets within a reasonable range (e.g., a router has a set of legitimate IP addresses for its interfaces) to increase the chances of our test packets to take all available paths. The original source address can be recorded in the test payload to allow the destination to identify the source of the test packets.

We have described the proposed active probing methodology in terms of probing schedule, the number of test packets for a certain accuracy, the test packet type and the packet size. With regard to a control protocol to initiate and to maintain monitoring sessions between endpoints, the existing protocols such as Cisco SAA (Service Assurance Agent) [16][5] or IPPM one-way active measurement protocol (OWAMP) [13] can be used with little modification.

### 5.5.2 Validation

To validate the proposed technique, we emulate active test packets in the following manner[6]. Given an estimation interval ($I$) and accuracy parameters ($\{\varepsilon, \eta\}$), whenever the random timer ($t$) expires, we choose the next arriving packet from the data sets, and use its delay as an active test packet measurement. The accuracy parameters are set to be $\varepsilon = 10\mu s$[7] and $\eta = 0.05$ to estimate $.99^{th}$ quantile of delay. We have used 0.001 and 0.0005 for $f(q_p)$. The computed numbers of samples to ensure the estimation accuracy are only 423 and 1526, respectively.

The estimated $.99^{th}$ quantiles over 10 minute intervals using 423 packets are compared with the actual $.99^{th}$ quantiles in Figure 5.17. Using the same number

---

[5] SAA (Service Assurance Agent) is an active probing facility implemented in Cisco routers to enable network performance measurement.

[6] We could not perform probing simultaneously to passive trace collection since all long-haul links on the Sprint backbone have been upgraded to OC-192 after the trace collection.

[7] This small error bound is chosen to show the feasibility of the proposed sampling.

of 423 test packets, the estimated quantiles are compared with the actual ones over 30 minute interval in Figure 5.18. Using such small numbers of packets, the estimated quantiles are very close to the actual ones, for all the data sets and estimation intervals.

To assess the statistical accuracy, we conduct experiments over an estimation interval (30 minutes) as many as 500 times. For $0.99^{th}$ quantile ($q_{.99}$), we desire the error to be less than $\varepsilon$ with probability of $1 - \eta$. We compare the estimated quantile from each experiment with the actual quantile from the total passive measurements. Figure 5.19(a) displays the estimation error in each experiment. Most errors are less then $10\mu s$ which is the error bound $\varepsilon$. To validate the statistical guarantee of accuracy, in Figure 5.19(b), we plot the cumulative empirical probability of errors in quantile estimation. The $y$ axis is the experimental cumulative probability that the estimate error is less than $x$. It illustrates that indeed 95% of the experiments give estimation error of less than $10\mu s$, which conforms to the pre-specified accuracy parameters.

Another key metric for the performance of a sampling technique is the variance of an estimator. Small variance in estimation is a desired feature for any sampling method, as it tells the estimate is more reliable. In the previous section, we have shown that the proposed scheme enables us to bound the variance of the estimates in terms of the accuracy parameters, i.e. $1/z_p = \left( \frac{\varepsilon}{\Phi^{-1}(1-\eta/2)} \right)^2$. Table 5.2 shows the variance of the estimates from the proposed scheme. The variances are indeed bounded by the value given in Eq. (5.9) given in Section 5.4.

**Table 5.2** Bounded variance of estimates ($\{\varepsilon, \eta\} = \{10\mu s, 0.05\}, p = 0.99$)

| $1/z_p$ | Data Set | 1 | 2 | 3 |
|---------|----------|-----|-----|-----|
| 25.95 | $Var(\hat{q}_p)$ | 11.97 | 25.72 | 25.55 |

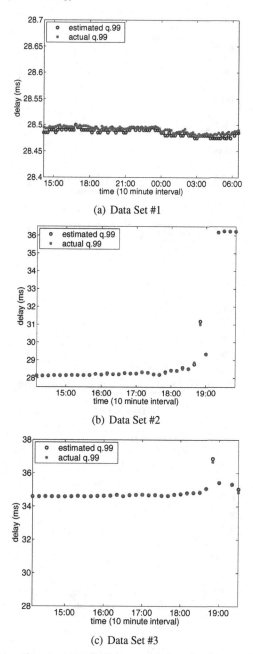

**Fig. 5.17** Actual and estimated $.99^{th}$ quantiles (10 minute estimation interval)

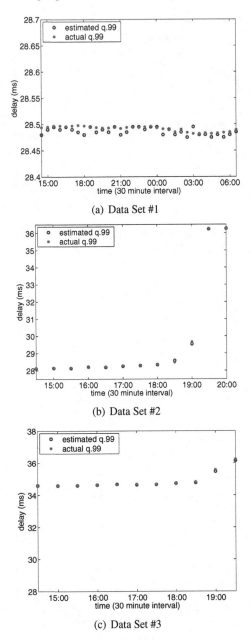

(a) Data Set #1

(b) Data Set #2

(c) Data Set #3

**Fig. 5.18** Actual and estimated .99$^{th}$ quantiles (30 minute estimation interval)

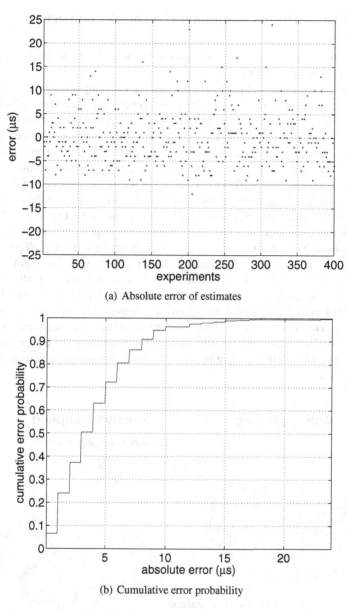

(a)  Absolute error of estimates

(b)  Cumulative error probability

**Fig. 5.19** Quantile estimation with bounded error ($\{\varepsilon, \eta\} = \{10\mu s, 0.05\}, p = 0.99$)

## 5.6 Active vs. Passive Sampling: Overhead Comparison

In this section, we compare a network-wide overhead of our active measurement method to a passive sampling technique.

For the comparison, we first describe a passive sampling process for delay measurement in a network (See Figure 5.20 for reference). We sketch here a hash based scheme proposed in [2]. For delay measurement, all regular packets are hashed and passively sampled based on their hash values and time-stamped at the measurement points. To capture the same sets of packets on different measurement points, the same hash function is used to sample packets at all measurement points. Then, the collected packets are exported to a central server where the same pair of packets are identified and the delay is computed. In order to reduce the bandwidth consumed when exporting those samples, only a hash of the packet ID is exported, rather than the whole packet header. The downside of this technique is to increase the risk of packet mismatch at a central sever. The central server then matches all packets and computes the delay from the difference of their timestamps. The method can be optimized using routing information in order to ease the task of finding pairs of measurement points where packets might have traversed from one to the other.

Note that even with passive measurement, measured packets should be transferred to a central server to combine time information from measurement points, since for one delay value, *two* measurement points are involved, i.e., the source and the sink. Therefore, either active or passive, delay measurement consumes bandwidth by nature.

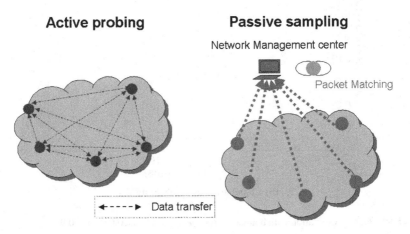

**Fig. 5.20** Active and passive delay measurements.

In order to compare the overhead, we consider the case where the number of delay samples are equal so as to achieve the same accuracy of estimation from both methods. Assuming the number of samples is small as shown in Section 5.4, the performance of regular traffic would not be affected by measurements for both active and passive measurement.

We ignore the control protocol overhead for signaling among routers (active probing) or between routers and a central server (passive sampling and active probing[8]), which we expect to be similar in both methods. In addition, both active and passive monitoring systems can be either implemented as an integral part of routers in an embedded manner [16] or as a stand-alone out-of-router measurement system.

First, let us analyze the bandwidth used by measurement data. Consider the number of bytes used to report one packet delay. In a passive method, we transfer only the packet identifier (hash value) rather than the entire packet header and payload. Each packet hash value should be transferred with its *timestamp*. Then, in order to compute one delay from one point to the other, two packets are required with a passive sampling. Suppose $n^*$ packets are required for a given accuracy. For a given pair of measurement points, the number of packets that have to be sent is $n^*$ with the active method. For the passive method, note that only a *portion* of the packets retrieved at a source router are sent to the sink router of the measurement interest. Similarly, only a portion, the so called traffic *fanout factor* $P$, (where $0 \leq P \leq 1$) of the packets at the sink router is originated from the source router. Therefore, in order to produce the needed number of delay samples, the number of measured packets has to be scaled accordingly. For example, let $P_{A,B}$ be the portion of total traffic from a measurement point $A$ to $B$. Then the number of samples at link $A$ should be scaled up by $P_{A,B}$ to produce the required number of matched packets on average. Similarly, the number of samples at link $B$ should be scaled up by a factor of $P_{B,A}$. Thus, the number of samples for a pair of delay measurements with the passive sampling ($n_{pass}^{op}$) is

$$n_{pass}^{op} = \frac{n^*}{P_{A,B}} + \frac{n^*}{P_{B,A}} \tag{5.10}$$

Now, we consider a network-wide number of samples in an ISP. With the passive measurement, the number of samples increases linearly with the number of measurement points, say $N_{mp}$, i.e., $\frac{2}{\bar{p}} \cdot n^* \cdot N_{mp}$. Meanwhile, the traffic fanout factor becomes inversely proportional to the number of measurement points. Let us denote the network-wide number of samples with a passive sampling as $n_{pass}^{nw}$. Then, for a network with $N_{mp}$ number of measurement points and an average fanout factor $P_{avg}$, $n_{pass}^{nw}$. The computation is

$$n_{pass}^{nw} = \sum_{(i,j)\in\{pairs\}} \left( \frac{n^*}{P_{i,j}} + \frac{n^*}{P_{j,i}} \right) \tag{5.11}$$

$$\approx \frac{n^*}{P_{avg}} \cdot N_{mp} \approx n^* N_{mp}^2 \tag{5.12}$$

---

[8] This additional signaling is required in active probing to report the estimated quantiles to a central server.

where we approximated the average fanout factor with the inverse of total number of measurement points in the network (i.e., $P_{avg} \approx 1/N_{mp}$).

In the active measurement, the number of samples grows linearly with the number of pair of measurement points, $N_{pairs}$, or quadratically with the number of measurement points. We denote the network-wide number of samples with the active measurement as $n_{actv}^{nw}$, and it is computed as below:

$$n_{actv}^{nw} = n^* \cdot N_{pairs} = n^* \cdot N_{mp} \cdot (N_{mp} - 1) \tag{5.13}$$

$$\approx n^* \cdot N_{mp}^2 \tag{5.14}$$

To assess the amount of actual bandwidth consumed, let us assume 64 bytes are used for both an active test packet and a passive packet sample. For a passive packet sample, suppose 4 bytes for a packet hash value, 8 bytes for a timestamp, 4 bytes for a source router address, 4 bytes for a link identifier, and 20 bytes for an export protocol header (in order for a central server to recognize the measurement data) are used. Including 20 and 8 bytes for IP and UDP headers of the exported packet, it totals 64 bytes for one packet data. For a measurement interval, we also assume 1000 samples are used. Figure 5.21(a) illustrates the bandwidth usage for the two schemes with varying numbers of measurement points. The advantage of the active method is prominent when a small number of measurement points are measured. If most of the measurement points are measured in a network, the number of samples from both methods becomes similar. In practice, traffic fanout exhibits a large disparity among measurement points[9]. In addition, the fanout factor is not known in advance and varies over time, making it hard to ensure the number of samples in passive sampling. Furthermore, there may be very little or no match between the packets sampled at two measurement points. In that case, it may not be possible for a passive measurement to produce enough samples to obtain a delay estimate with any reasonable accuracy. On the other hand, the active method injects only a fixed, required number of samples, regardless of the traffic load between measurement points, thus ensuring the accuracy of the measurement. Therefore, when a portion of measurement pairs are measured, the passive sampling consumes more bandwidth, and renders itself more intrusive than the active measurement. In addition, passive measurement requires all packets to be hashed, potentially affecting the performance of the forwarding path of the measurement point.

Now we consider memory requirement either at a router or at a central station. In a passive measurement, sets of transferred packets have to be kept at a central station for a long enough duration of packet delay within a network. Then a set of stored packets from a measurement point will be matched with ones from another measurement point for delay computation. On the other hand, in the active measurement, each measurement point computes the delay of a test packet on its arrival at its destination. Thus, only the data relevant to delay statistics (e.g., histogram) needs to be kept at a sink router. Figure 5.21(b) compares memory requirement of the two

---

[9] An instance of a PoP level traffic matrix showed that the fanout varies from 0.001% to 40% in the network of our study. The fanout factor would dramatically decrease in router or link level [11].

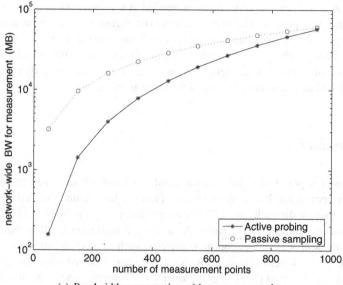

(a) Bandwidth consumption with measurement data

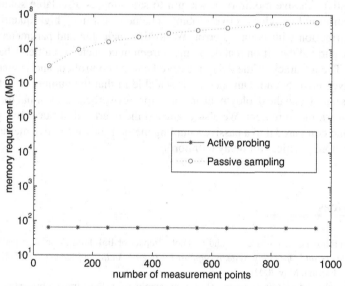

(b) Memory requirement at a measurement point or a central station

**Fig. 5.21** Active probing vs. passive sampling: bandwidth and memory usage comparison.

schemes. A fixed amount of memory is needed at a measurement point in the active scheme. In a passive measurement, however, memory requirement at a central station larger than active scheme and grows with the network size.

Taking bandwidth consumption and memory requirement into consideration, for the purpose of network wide delay monitoring, we find that an active probing is more practical and less intrusive over a passive sampling.

## 5.7 Summary

We proposed a practical delay measurement methodology designed to be implemented in operational backbone networks. First we have found out that the packet size, queueing on the path, and ECMPs weigh differently on the shape of the delay distribution, depending on the paths. As a delay distribution is often multi-modal and extreme delays are not uncommon, metrics, such as median and average, do not reflect the network performance most packets experience. We identify high quantiles (between 0.95 and 0.99) of delay over 10-30 minute time intervalas the right metric that captures the network delay performance in a meaningful way. Instead of exhaustive, passive monitoring, we turn to sampling as a scalable solution and propose an pseudo random active probing scheme to estimate high quantiles of a delay distribution with bounded error. We justify each step and parameters of the technique and validate it on real delay measurement collected on a tier-1 backbone network. The accuracy of the delay measured can be controlled, and is guaranteed with a given error bound. Our method is scalable in that the number of active test packets is small, and the deployment and monitoring overhead is minimal for a backbone network measurement. We also evaluated the overhead of our active probing scheme and compared it to a passive sampling method showing active measurement becomes more practical for delay monitoring.

## References

1. C. Boutremans, G. Iannaccone, and C. Diot. Impact of link failures on voip performance. In *Network and Operating Systems Support for Digital Audio and Video (NOSSDAV)*, Miami Beach, Florida, May 2002.
2. N. Duffield and M. Grossglauser. Trajectory sampling for direct traffic observation. In *Proceedings of ACM SIGCOMM*, 2000.
3. C. Fraleigh, S. Moon, B. Lyles, C. Cotton, M. Khan, D. Moll, R. Rockell, T. Seely, and C. Diot. Packet-level traffic measurements from the Sprint IP backbone. *To be published in IEEE Network*, 2003.
4. U. Hengartner, S. Moon, R. Mortier, and C. Diot. Detection and analysis of routing loops in packet traces. In *ACM SIGCOMM Internet Measurement Workshop*, Marseille, France, November 2002.

5. G. Huston. *ISP Survival Guide: Strategies for Running a Competitive ISP*. John Wiley & Sons, October 1998.

6. G. Iannaccone, C-N. Chuah, R. Mortier, S. Bhattacharyya, and C. Diot. Analysis of link failures over an ip backbone. In *ACM SIGCOMM Internet Measurement Workshop*, Marseille, France, November 2002.

7. Sprint ATL IPMon project. *http://ipmon.sprint.com*.

8. R. Jain. *The Art of Computer Systems Performance Analysis: Techniques for Experimental Design, Measurement, Simulation, and Modeling*. Wiley-Interscience, April 1991.

9. S. Jamin, C. Jin, Y. Jin, R. Raz, Y. Shavitt, and L. Zhang. On the placement of Internet Instrumentation. In *Proceedings of INFOCOM*, Tel Aviv, Israel, March 2000.

10. A. Markopoulou, G. Iannaccone, S. Bhattacharyya, C-N. Chuah, and C. Diot. Characterization of failures in an IP backbone. In *IEEE Infocom*, Hong Kong, November 2004.

11. A. Medina, N. Taft, K. Salamatian, S. Bhattacharyya, and C. Diot. Traffic matrix estimation: Existing techniques and new directions. In *Proceedings of ACM SIGCOMM*, Pittsburgh, August 2002.

12. D. Moore. *The Basic Practice of Statistics, 3rd ed.* W. H. Freeman Publishers, April 2003.

13. OWAMP. IETF IPPM draft: A One-Way Active Measurement Protocol. *http://www.ietf.org/internet-drafts/draft-ietf-ippm-owdp-07.txt*.

14. K. Papagiannaki, S. Moon, C. Fraleigh, P. Thiran, and C. Diot. Measurement and analysis of single-hop delay on an IP backbone network. In *Proceedings of INFOCOM*, San Francisco, CA, April 2002.

15. V. Paxson, A.K. Adams, and M. Mathis. Experiences with nimi. In *Proceedings of Passive and Active Measurement Workshop*, Hamilton, New Zealand, April 2000.

16. SAA. Cisco Service Assurance Agent. *http://www.cisco.com*.

17. R. Serfling. *Approximation Theorems of Mathematical Statistics*. Wiley, 1980.

18. *PSLA Management Handbook*. April 2002.

19. D. Thaler and C. Hopps. Multipath issues in unicast and multicast next-hop selection. Internet Engineering Task Force Request for Comments: 2991, November 2000.

20. Yin Zhang, Nick Duffield, Vern Paxson, and Scott Shenker. On the Constancy of Internet Path Properties. In *ACM SIGCOMM Internet Measurement Workshop*, San Francisco, California, USA, November 2001.

# Chapter 6
# Concluding Remarks

## 6.1 Summary

In this book, we have analyzed network traffic and performance parameters, and have developed scalable schemes to monitor them accurately. We consider the measurement problems of total traffic load, flow volume, and point-to-point delay.

On total load measurement, we have shown that the minimum number of samples needed to maintain the prescribed accuracy is proportional to the squared coefficient of the variation (*SCV*) of packet size distribution. Since we do not have *a priori* knowledge about key traffic parameters – *SCV* of packet size distribution and the number of packets – these parameters are predicted using AR model. The sampling probability is then determined based on these predicted parameters and varies adaptively according to traffic dynamics. From the sampled packets, the traffic load is then estimated. We have also derived a theoretical upper bound on the variance of estimation error that affects the robustness of a change detection algorithm. We have experimented with real traffic traces and demonstrated that the proposed adaptive random sampling is very effective in that it achieves the desired accuracy, while also yielding significant reduction in the fraction of sampled data. As part of our ongoing efforts, we are working on extending the proposed sampling technique to address the problem of flow size estimation.

The problem of flow measurement involves more complicated issues, such as diversity of flow size, dynamic flow arrival and duration, and time-varying rate of flows. Since a small percentage of flows are observed to account for a large percentage of the total traffic, we focused on the accurate measurement of elephant flows. We define an elephant flow in such a way that captures high packet count, high byte count as well as high rate flows. We enhanced the adaptive sampling method that adjusts the sampling rate so as to bound the error in flow volume estimation without excessive oversampling. The proposed method based on stratified random sampling divides time into strata, and within each stratum, samples packets randomly at a rate determined according to the minimum number of samples needed to achieve the desired accuracy. The technique can be applied to any granularity of flow definition.

In order to tackle the issue of delay measurement, we first analyze point-to-point delay from an operational tier-1 backbone network. We developed methods to isolate factors constituting point-to-point delay. We have isolated the fixed components in delay, namely, propagation delay, transmission delay, and per-packet overhead at the router. When there is more than one path between two points, we have determined the path of each packet, using the minimum delay per 2-tuple flow and the TTL delta. Once we have identified a set of packets that followed the same path, we obtain the minimum path transit time per packet size and subtract it from point-to-point delay to obtain only the variable delay. When the link utilization on all links of the path is under 90%, the 99th percentile of the variable delay remains insignificant most of time. However, when a link on the path is utilized over 90%, it becomes a bottleneck in the sense that the weight of the variable delay distribution shifts and even the 90th percentile of the variable delay reaches above 500 $\mu$s and the 99th percentile, well beyond 1 ms. Though rare and few in numbers, there are peaks in variable delay, of which the magnitude reaches tens of milliseconds. We observe such peaks even below 90% of the bottleneck link utilization, and they do affect the tail shape of the distribution. To the best of our knowledge, it is the first study that shows real network packet delays.

We have found out that shapes of delay distributions are quite different. It is because the packet size, queueing on the path, and ECMPs weigh differently on the shape of the delay distribution, depending on the path characteristics. As a result, a delay distribution is often multi-modal and extreme delays are not uncommon, and metrics, such as median and average, do not reflect the network performance most packets experience. We identify high quantiles as the right metric that captures the network delay performance in a meaningful way. Instead of exhaustive, passive monitoring, we turn to sampling as a scalable solution and propose an active sampling scheme to estimate high quantiles of a delay distribution with bounded error. With a relatively small number of probes, we can estimate high quantiles with a guaranteed level of accuracy.

Our research in this book shows that as a *scalable* Internet monitoring methodology, packet *sampling* can be used to estimate network performance and traffic *accurately*.

## 6.2 Future Directions

The following are some future research directions related to the contributions of this book.

### 6.2.1 Network Jitter Analysis and Measurement Scheme

Jointly with delay, jitter is another important network performance parameter. Jitter has a direct impact on the user perceived performance with streaming applications. End host systems leverage buffers to extenuate jerkiness of applications due to jitter. Thus, analysis of jitter performance gives insights on optimal buffer provisioning. From an ISP's point of view, it is a viable approach to enhance the current SLA with jitter performance. Understanding network jitter behavior and what can be measured and offered as SLA is a prerequisite to provisioning SLA. Measuring and analyzing network jitter in a backbone would be a first step in order to find practical definitions and metrics of jitter. Then, designing and developing a scalable measurement methodology is the useful next step to monitor jitter for the assessment of performance of streaming applications.

### 6.2.2 Traffic Matrix from Packet Level Traffic Information

Traffic matrix is a representation of traffic exchanges between nodes in a network. It can help greatly from the diagnosis and management of network congestion to network design and capacity planning. Particularly, prefix-level (flow) traffic matrices give answers to the following important questions. How much traffic goes to/from an ISP to/from certain regions? What type of applications are used in the network? Furthermore, prefix-level traffic matrices make ground for the higher granularity of traffic matrices. However, current traffic matrix estimation techniques are limited to large granularity such as pop-level or router level. This is due to the huge scale of measurement data. The goal of the research is to compute fast and efficient fine-grain traffic matrices by an optimal sampling technique and to observe the variability/dynamics of the network traffic that would be otherwise impossible.

### 6.2.3 Optimal Placement of Measurement Systems and Inference of a Network Performance

In addition to monitoring the aforementioned temporal fluctuation of traffic at a link/router, understanding the spatial flow of traffic is important to engineer the network performance, such as route optimization or planning of fail-over strategies. To have better spatial traffic information in a large network with given cost and scalability constraints, a careful choice of the measurement instrumentation placement is essential. Where should we place the measurement-box in the network under the constraints? Given the limited measurement places, how much can we infer from the measurement? To address these issues, first we plan to identify and classify a variety of usages of traffic measurements. Formal definitions of given objectives have to be followed. Optimization techniques will be used for the solution of the formally defined measurement placement problem. Developing an inference/estimation technique is an important last step in speculating about the whole network, as well as evaluating the performance of the measurement.